Lab Manual

to accompany

Agricultural Mechanics
Fundamentals & Applications

4th Edition

Ray V. Herren
Elmer Cooper
William Hamilton

**Delmar is proud
to support FFA activities.**

Join us on the web at

Lab Manual

to accompany

Agricultural Mechanics
Fundamentals & Applications

4th Edition

Ray V. Herren
Elmer Cooper
William Hamilton

DELMAR
THOMSON LEARNING

Australia Canada Mexico Singapore Spain United Kingdom United States

COPYRIGHT © 2002 by Delmar, a division of Thomson Learning, Inc. Thomson Learning™ is a trademark used herein under license.

Printed in the United States
5 XXX 05 04 03

For more information contact Delmar,
3 Columbia Circle, PO Box 15015, Albany, NY 12212-0515.
Or find us on the World Wide Web at http://www.delmar.com

ALL RIGHTS RESERVED. No part of this work covered by the copyright hereon may be reproduced or used in any form or by any means—graphic, electronic, or mechanical, including photocopying, recording, taping, Web distribution or information storage and retrieval systems—without written permission of the publisher.

For permission to use material from this text or product, contact us by
Tel (800) 730-2214
Fax (800) 730-2215
www.thomsonrights.com

Library of Congress Cataloging Number: 00-030671

ISBN: 0-7668-1413-0

NOTICE TO THE READER

Publisher does not warrant or guarantee any of the products described herein or perform any independent analysis in connection with any of the product information contained herein. Publisher does not assume, and expressly disclaims, any obligation to obtain and include information other than that provided to it by the manufacturer.

The reader is expressly warned to consider and adopt all safety precautions that might be indicated by the activities herein and to avoid all potential hazards. By following the instructions contained herein, the reader willingly assumes all risks in connection with such instructions.

The Publisher makes no representation or warranties of any kind, including but not limited to, the warranties of fitness for particular purpose or merchantability, nor are any such representations implied with respect to the material set forth herein, and the publisher takes no responsibility with respect to such material. The publisher shall not be liable for any special, consequential, or exemplary damages resulting, in whole or part, from the readers' use of, or reliance upon, this material.

CONTENTS

SECTION	1	EXPLORING CAREERS IN AGRICULTURAL MECHANICS	1
Unit	1	Mechanics in the World of Agriculture	1
Unit	2	Career Options in Agricultural Mechanics	3
SECTION	2	USING THE AGRICULTURAL MECHANICS SHOP	5
Unit	3	Shop Orientation and Procedures	5
Unit	4	Personal Safety in Agricultural Mechanics	10
Unit	5	Reducing Hazards in Agricultural Mechanics	15
Unit	6	Shop Cleanup and Organization	26
SECTION	3	HAND WOODWORKING AND METALWORKING	28
Unit	7	Hand Tools, Fasteners, and Hardware	28
Unit	8	Layout Tools and Procedures	34
Unit	9	Selecting, Cutting, and Shaping Wood	43
Unit	10	Fastening Wood	51
Unit	11	Finishing Wood	59
Unit	12	Identifying, Marking, Cutting, and Bending Metal	68
Unit	13	Fastening Metal	80
SECTION	4	POWER TOOLS IN THE AGRICULTURAL MECHANICS SHOP	93
Unit	14	Portable Power Tools	93
Unit	15	Woodworking with Power Machines	103
Unit	16	Metalworking with Power Machines	115
SECTION	5	PROJECT PLANNING	122
Unit	17	Sketching and Drawing Projects	122
Unit	18	Figuring a Bill of Materials	128
Unit	19	Selecting, Planning, and Building a Project	133
SECTION	6	TOOL FITTING	142
Unit	20	Repairing and Reconditioning Tools	142
Unit	21	Sharpening Tools	150
SECTION	7	GAS HEATING, CUTTING, BRAZING, AND WELDING	162
Unit	22	Using Gas Welding Equipment	162
Unit	23	Cutting with Oxyfuels	169
Unit	24	Brazing and Welding with Oxyacetylene	172
SECTION	8	ARC WELDING	183
Unit	25	Selecting and Using Arc Welding Equipment	183
Unit	26	Arc Welding Mild Steel and MIG/TIG Welding	188

SECTION	9	PAINTING	203
Unit	27	Preparing Wood and Metal for Painting	203
Unit	28	Selecting and Applying Painting Materials	213
SECTION	**10**	**SMALL GAS ENGINES**	**222**
Unit	29	Fundamentals of Small Engines	222
Unit	30	Small Engine Maintenance and Repair	226
SECTION	**11**	**ELECTRICITY AND ELECTRONICS**	**239**
Unit	31	Electrical Principles and Wiring Materials	239
Unit	32	Installing Branch Circuits	249
Unit	33	Electronics in Agriculture	256
Unit	34	Electric Motors, Drives, and Controls	260
SECTION	**12**	**PLUMBING, HYDRAULIC, AND PNEUMATIC SYSTEMS**	**266**
Unit	35	Plumbing	266
Unit	36	Irrigation Technology	273
Unit	37	Hydraulic, Pneumatic, and Robotic Power	276
SECTION	**13**	**CONCRETE AND MASONRY**	**283**
Unit	38	Concrete and Masonry	283
SECTION	**14**	**AGRICULTURAL STRUCTURES**	**292**
Unit	39	Planning and Constructing Agricultural Structures	292
Unit	40	Aquaculture, Greenhouse, and Hydroponics Structures	304

LIST OF FIGURES

Figure 7-1. Types of Nails 31

Figure 7-2. Screw Classifications 31

Figure 7-3. Common Bolts 32

Figure 7-4. Types of Washers 32

Figure 7-5. Types of Hinges 33

Figure 7-6. Flush Plates 33

Figure 8-1. Inch Ruler Readings 35

Figure 8-2. Metric Ruler Readings 35

Figure 8-3. Layout Tools 36

Figure 10-1. Common Wood Joints 52

Figure 12-1. Identifying Files 72

Figure 12-2. Identifying Metal Snips 73

Figure 19-1. Agricultural Mechanics Competencies 134

Figure 22-1. Oxyacetylene Flames 168

Figure 26-1. Components of the MIG Welder 200

Figure 26-2. MIG Welding Defects 201

Figure 28-1. Parts of a Spray Gun 215

Figure 31-1. Electrical Symbols 245

Figure 38-1. Concrete Tools 285

Figure 38-2. Types of Concrete Blocks 286

SECTION 1 EXPLORING CAREERS IN AGRICULTURAL MECHANICS

Unit 1 Mechanics in the World of Agriculture

Many jobs and careers are available in the field of agricultural mechanics. This unit is designed to help you explore some of these.

Class Activity 1-1

Studying Agricultural Mechanics Careers

Job Sheet 1-1

Name _____ Date _____

Studying Agricultural Mechanics Careers

Objective

When you have completed this activity you will have demonstrated an understanding of agricultural mechanics careers.

Tools and Materials

- Job sheet 1-1
- Textbook
- Pen or pencil

Procedure

Complete the following questions.

1. What is an agricultural occupations cluster?
   _____
   _____

2. Which of the following would be included?

	Yes	No
Salesclerk	___	___
Production	___	___
Forestry	___	___
Tractor mechanic	___	___
Agricultural mechanics	___	___
Turf management	___	___
Set-up man	___	___
Landscape design	___	___
Delivery man	___	___

3. How many people does our text suggest the average farmer may feed _____, top farmer? _____

4. Define profit. _____
   _____
   _____

5. For every worker in production agriculture there are probably _____ others working in a related area.

6. Mechanics who use high technology are called _____.

7. Why is agriculture called a basic industry?
   _____
   _____
   _____

8. The role of agricultural mechanics in America's productivity is: major or minor? (Circle the best answer.)

9. List five jobs requiring agricultural mechanics skills.
   _____
   _____
   _____
   _____
   _____

10. Why was John Deere's improvement of Thomas Jeffersons's iron plow so important?
    _____
    _____

11. Why was Whitney's cotton gin so important in the 1790s?
    _____
    _____

12. Many other mechanical inventions were important because the improved
    _____

13. Why are American machines not widely used in other countries?
    _____
    _____

14. Can you name on use of computers in agriculture?
    _____

15. How may the use of computers and robotics influence agriculture in the future?
    _____
    _____
    _____

Instructor's Score or Approval _____

Unit 2 Career Options in Agricultural Mechanics

Unit 2 continues the study of agricultural occupations.

Class Activity 2-1

Agricultural Occupations

Job Sheet 2-1

Name _____ Date _____

Agricultural Occupations

Objective

When you have completed this activity you will have demonstrated an understanding of agricultural occupations.

Tools and Materials

Job sheet 2-1

Textbook

Pen or pencil

Procedure

Complete the following questions.

1. Off-farm agricultural jobs are those requiring _____ _____ but are not farming or ranching.

2. What attracts many people to careers in agriculture?
   _____

3. Americans use approximately _____ percent of their income for food.

4. How does this percentage compare to other parts of the world?
   _____

5. What are the levels of employment in agriculture?
   _____

6. What percent of employment in agriculture is in production?
   _____

7. What education is required for the professional level?
   _____
   _____

8. List the eight divisions of agribusiness and agricultural production.
   _____
   _____
   _____
   _____
   _____

9. List the six divisions in agricultural science.

   _____
   _____
   _____
   _____

10. List the seven divisions in renewable natural resources.

    _____
    _____
    _____
    _____

11. What are the specific job title categories in agricultural mechanics?

    _____
    _____
    _____
    _____

12. List some questions you should ask yourself when looking into career choices.

    _____
    _____
    _____
    _____

13. How can you prepare yourself for a career in the field of agriculture?

    _____
    _____
    _____
    _____
    _____

14. Which student organization can help you develop useful skills for agricultural or other occupations?

    _____

Instructor's Score or Approval _____

Section 2 Using the agricultural mechanics shop

Unit 3 Shop Orientation and Procedures

The shop will offer you many hands on opportunities to learn skills useful to you in many of the careers discussed in the first two units. If you are to get the most from your experiences, you will need to become familiar with the shop, tools, and safety.

The activities in this unit are designed to aid you in identifying the capabilities of your shop and becoming acquainted with its tools and work areas. Complete each activity with the idea of gaining the most knowledge you can from your shop experience.

Class Activity 3-1

Identifying the Importance of Agricultural Mechanics Activities

Shop Activity 3-2

Sketching and Labeling the Agricultural Mechanics Shop

Shop Activity 3-3

Completing the Shop Policies and Procedures Form

Class Activity 3-4

Completing the Allergies and Physical Problems Statement

Job Sheet 3-1

Name _____ Date _____

Identifying the Importance of Agricultural Mechanics Activities

Objective

Upon the completion of this activity you will have demonstrated the ability to identify the importance of agricultural mechanics activities.

Tools and Materials

Job sheet 3-1

Pen or pencil

Procedure

Using the text as a reference, complete the following job sheet.

1. Complete the following statement: The agricultural mechanics shop is an important learning place because

_____
_____
_____
_____
_____

2. List the categories of operations given in the text.

   _____
   _____
   _____
   _____

3. Why are open areas important in the shop?

   _____
   _____
   _____
   _____

4. Why are electrical skills important in agricultural mechanics?

   _____
   _____

5. What are the agricultural power and machinery experiences named in the text?

   _____
   _____
   _____

6. The agricultural mechanics shop should be adequate in size for safe and efficient work by _____ high school students.

7. Tell why you agree or disagree with the shop policies and procedures in Figure 3-15.

   _____
   _____
   _____
   _____
   _____
   _____
   _____
   _____

Instructor's Score or Approval _____

> **Job Sheet 3-2**

Name _____ Date _____

Sketching and Labeling the Agricultural Mechanics Shop

Objective

Upon the completion of this activity you will have demonstrated the ability to identify and sketch the major work areas, stationary power tools, and the shop safety features.

Tools and Materials

Job sheet 3-2

Drawing board

T-square

Triangles

Pencils

Drawing paper

100-foot tape

Procedure

1. Measure shop dimensions and draw the shop outline.
2. Sketch in the stationary tools.
3. Identify the work areas in the shop.
4. Identify the safety features in the shop.
5. Label the power tools.
6. Label the work areas.
7. Label the storage areas.
8. Label the safety features.

Instructor's Score or Approval _____

Job Sheet 3-3

Name _____ Date _____

Completing the Shop Policies and Procedures Form

Objective

Upon completing this activity you will have demonstrated the acceptance of the shop safety policy.

Tools and Materials

Job sheet 3-3

Copy of the policies, Figure 3-15

Pen or pencil

Procedure

1. Obtain a copy from the instructor.
2. Discuss the policy and procedure form in class.
3. Discuss the form with my parents or guardian.
4. Have the form signed by parents or guardian.
5. Return the signed form to my instructor.

Instructor's Score or Approval _____

Job Sheet 3-4

Name _____ Date _____

Completing the Allergies and Physical Problems Statement

Objective

Upon completing this activity you will have provided your teacher with any allergic or physical problems that could affect your work in the shop.

Tools and Materials

    Job sheet 3-4

    Copy of Figure 3-16

    Pen or pencil

Procedure

1. Obtain a copy from the instructor.
2. Discuss the form in class.
3. Discuss the form with your parents or guardian.
4. Have the form signed by parents or guardian.
5. Return the form to your instructor.

Instructor's Score or Approval _____

Unit 4 Personal Safety in Agricultural Mechanics

Most injuries are avoidable. No one plans to be injured, and most people dislike being reminded of safety precautions yet injuries occur. In the last unit you completed a shop policy form with emphasis on safety. In this unit you will look at specific personal safety items.

Class Activity 4-1

Identifying Safety in the Agricultural Mechanics Shop

Class Activity 4-2

Identifying the Meaning of Shop Color Codes

Shop Activity 4-3

Identifying Shop Color Codes

Class Activity 4-4

Identifying Protective Clothing

Job Sheet 4-1

Name _____ Date _____

Identifying Safety in the Agricultural Mechanics Shop

Objective

Upon completing this activity you will have demonstrated the ability to identify safety in the agricultural mechanics shop.

Tools and Materials

Job sheet 4-1

Work sheet

Pen or pencil

Procedure

Using the text as a reference, complete the following work sheet.

1. Define safety.
   _____
   _____

2. How important is it to emphasize safety in the agricultural mechanics shop?
   _____
   _____
   _____

3. What are the common causes of accidents listed?
   _____
   _____
   _____
   _____
   _____

4. List the precautions needed to provide a safe workplace.

   _____
   _____
   _____
   _____
   _____
   _____
   _____
   _____
   _____
   _____
   _____

5. The best protection from injury is _____

6. The second best approach is
   _____
   _____

7. When are goggles or face shields required in the agricultural mechanics shop?
   _____
   _____

8. How should long hair be restrained in the shop?
   _____
   _____

9. What protective clothing should be used in the shop?
   _____
   _____
   _____

10. What are the characteristics of noise that cause earplugs or earmuffs to be recommended for shop use?
    _____
    _____

11. What is the decibel (dB) level OSHA has established as the maximum allowable? _____ dB

12. Why should every shop have an eye wash station?
    _____
    _____
    _____
    _____

Instructor's Score or Approval _____

Job Sheet 4-2

Name _____ Date _____

Identifying the Meaning of Shop Color Codes

Objective

Upon completing this activity you will have demonstrated the ability to identify the meaning of shop color codes.

Tools and Materials

    Job sheet 4-2

    Pen or pencil

Procedure

Match the colors in column I with their meaning in column II.

Column I

    ___ 1. Red
    ___ 2. Orange
    ___ 3. Yellow
    ___ 4. Blue
    ___ 5. Green
    ___ 6. Black & yellow diagonal stripes
    ___ 7. White
    ___ 8. White & black stripes
    ___ 9. Gray

Column II

    a. warning
    b. work areas
    c. traffic areas
    d. danger
    e. caution
    f. information
    g. safety
    h. radio activity
    i. work area markings

Instructor's Score or Approval _____

Job Sheet 4-3

Name _____ Date _____

Identifying Shop Color Codes

Objective

Upon completing this activity you will have demonstrated the ability to identify the safety color codes.

Tools and Materials

    Job sheet 4-3

    A copy of the shop color code

    Pen or pencil

Procedure

    Has your agricultural mechanics shop been partially or fully color coded?   ___ Yes ___ No

If Yes, proceed, if No, go to Job sheet 4-4, Identifying Protective Clothing.

1. Safety switches are marked in red.   ___ Yes ___ No
2. Fire equipment is marked with red.   ___ Yes ___ No
3. Machine hazards are marked in orange.   ___ Yes ___ No
4. Backgrounds of switches are orange.   ___ Yes ___ No
5. Backgrounds for levers and controls are orange.   ___ Yes ___ No
6. Machine controls or adjustments are marked in yellow.   ___ Yes ___ No
7. Stationary hazards such as stairs or protruding objects are marked in yellow and black stripes.   ___ Yes ___ No
8. Any information signs are in blue.   ___ Yes ___ No
9. Traffic areas are marked in white or yellow.   ___ Yes ___ No
10. Machine bodies are marked in gray or vista green.   ___ Yes ___ No
11. Waste containers are marked in aluminum.   ___ Yes ___ No
12. Ivory is used to improve visibility.   ___ Yes ___ No

Instructor's Score or Approval _____

Job Sheet 4-4

Name _____ Date _____

Identifying Protective Clothing

Objective

Upon completing this activity you will have demonstrated the ability to identify protective clothing.

Tools and Materials

Job sheet 4-4

Pen or pencil

Procedure

Match the items of protective clothing in column I with the related items in column II.

Column I		*Column II*	
____ 1.	Aprons	a.	hard hat or hairnet
____ 2.	Shop coat	b.	earplugs
____ 3.	Decibel	c.	bench work
____ 4.	Mask	d.	leather or rubber
____ 5.	Hair restraint	e.	dust
____ 6.	Hard hats	f.	eyes
____ 7.	Footwear	g.	frequent changes
____ 8.	Coveralls	h.	general body protection
____ 9.	Gloves	i.	dropped or flying objects
____ 10.	Face shields or goggles	j.	hands

Instructor's Score or Approval _____

Unit 5 Reducing Hazards in Agricultural Mechanics

A number of agricultural mechanics activities use processes that may cause fire if improperly attempted. This unit is geared to helping reduce the fire and other hazards in agricultural mechanics.

Read Unit 5 and refer to it as needed while completing the following activities.

Class Activity 5-1

Identifying Classes of Fires

Class Activity 5-2

Identifying the Components of the Fire Triangle

Class Activity 5-3

Identifying Appropriate Fire Extinguishers

Shop Activity 5-4

Identifying the Types and Locations of Fire Extinguishers in the Agricultural Mechanics Shop

Shop Activity 5-5

Doing a Monthly Fire Extinguisher Checklist

Class Activity 5-6

Identifying the Elements of a Chemical Label

Class Activity 5-7

Identifying the Procedure to Use in Case of an Accident

Class Activity 5-8

Identifying the Procedure to Use in Case of Fire

Class Activity 5-9

Identifying Requirements for the Use of the Slow Moving Vehicle (SMV) Emblem

Job Sheet 5-1

Name _____ Date _____

Identifying Classes of Fires

Objective

Upon the completion of this activity you will have demonstrated the ability to identify the classes of fires.

Tools and Materials

    Job sheet 5-1

    Pen or pencil

Procedure

Check the materials in the agricultural mechanics shop in the left-hand column and then match them with the class of fire hazard they represent in the right-hand column.

	Item	*Class*
____	Acetylene	____
____	Aluminum	____
____	Books and manuals	____
____	Electrical equipment	____
____	Gasoline	____
____	Grease	____
____	Lubricating oil	____
____	Lumber	____
____	Oxyfuels	____
____	Paints	____
____	Paper towels	____
____	Plywood	____
____	Protective clothing	____
____	Rope or twine	____
____	Sandpaper	____
____	Varnish	____
____	Welding curtains	____
____	Wooden benches	____
____	_____	____
____	_____	____
____	_____	____
____	_____	____

Instructor's Score or Approval _____

> **Job Sheet 5-2**

Name _____ Date _____

Identifying the Components of the Fire Triangle

Objective

Upon completing this activity you will have demonstrated the ability to identify the components of the fire triangle.

Tools and Materials

   Job sheet 5-2

   Pen or pencil

Procedure

   Complete the following worksheet.

1. What are the three components of the fire triangle?

   _____
   _____
   _____

2. What happens if one of the three is missing, or removed?

   _____
   _____

3. What methods are suggested for removing a part of the fire triangle?

   _____
   _____

Instructor's Score or Approval _____

Job Sheet 5-3

Name _____ Date _____

Identifying Appropriate Fire Extinguishers

Objective

Upon completing this activity you will have demonstrated the ability to identify appropriate fire extinguishers for the fire hazards in the shop.

Tools and Materials

    Job sheet 5-3

    Pen or pencil

Procedure

Using the checklist you completed in the first activity, write the class of fire extinguisher to use on a fire hazard with that source in the left column.

Class	Item
_____	Acetylene
_____	Aluminum
_____	Books and manuals
_____	Electrical equipment
_____	Gasoline
_____	Grease
_____	Lubricating oil
_____	Lumber
_____	Oxyfuels
_____	Paints
_____	Paper towels
_____	Plywood
_____	Protective clothing
_____	Rope or twine
_____	Sandpaper
_____	Varnish
_____	Welding curtains
_____	Wooden benches
_____	_____
_____	_____
_____	_____
_____	_____

Instructor's Score or Approval _____

Job Sheet 5-4

Name _____ Date _____

Identifying the Types and Locations of Fire Extinguishers in the Agricultural Mechanics Shop

Objective

Upon the completion of this activity you will have demonstrated the ability to identify types and locations of the fire extinguishers in the shop.

Tools and Materials

Job sheet 5-4

Pen or pencil

Type	Location
_____	_____
_____	_____
_____	_____
_____	_____
_____	_____
_____	_____
_____	_____
_____	_____

Instructor's Score or Approval _____

Job Sheet 5-5

Name _____ Date _____

Doing a Monthly Fire Extinguisher Checklist

Objective

Upon completing this activity you will have demonstrated the ability to do a fire extinguisher checklist.

Tools and Materials

Job sheet 5-5

Pen or pencil

Procedure

1. Obtain the location of the fire extinguisher you are to check from your instructor.
2. The fire extinguisher is located in a proper area.
3. The fire extinguisher is the right type for the area.
4. The fire extinguisher is free from visible damage.
5. Safety seals are in place.
6. It is free of dirt and rust.
7. The gauge or indicator is in the operable range.
8. It is of the proper weight.
9. The symbols and name plate are readable.
10. There are no nozzle obstructions.
11. When was the unit last serviced? _____

Instructor's Score or Approval _____

Job Sheet 5-6

Name _____ Date _____

Identifying the Elements of a Chemical Label

Objective

Upon the completion of this activity you will have demonstrated the ability to identify the elements of a chemical label.

Tools and Materials

    Job sheet 5-6

    Chemical label

    Pen or pencil

Procedure

Answer the following questions from the chemical label.

1. What is the use classification?

   _____

2. What is the brand name?

   _____

3. What is the common name?

   _____

4. What is the chemical name?

   _____

5. What is the formulation?

   _____

6. What are the ingredients?

   _____

7. What is the signal word?

   _____

8. What is the statement of practical treatment and antidote?

   _____

   _____

   _____

9. What are the directions for use?

   _____

   _____

   _____

10. What is the reentry statement?

    _____

11. What is the precautionary statement?

    _____

    _____

12. What are the storage and disposal directions?
   _____
   _____

13. What is the name and address of the manufacturer?
   _____
   _____
   _____

14. What is the EPA registration number?
   _____

15. What is the EPA establishment number?
   _____

16. What are the net contents?
   _____

Instructor's Score or Approval _____

Job Sheet 5-7

Name _____ Date _____

Identifying the Procedure to Use in Case of an Accident

Objective

Upon completing this activity you will have demonstrated the ability to identify the procedure to follow in case of an accident.

Tools and Materials

    Job sheet 5-7

    Pen or pencil

Procedure

Indicate with a Yes if the following actions should be taken and a No if they should not be taken. Incorrect items are included to test your knowledge of correct procedures.

_____ Avoid unnecessary movement of the victim.

_____ Call or send for help.

_____ Call qualified people for needed help.

_____ Check to see if the victim is breathing.

_____ Clear air passages if necessary.

_____ Control any bleeding.

_____ Cool the victim.

_____ Cover the victim.

_____ Get the victim on his or her feet.

_____ Handle the situation by yourself.

_____ Immobilize the victim if broken bones are suspected.

_____ Keep calm.

_____ Keep the victim lying down.

_____ Move the victim carefully in case of immediate danger such as fire.

_____ Move the victim to a more comfortable location.

_____ Notify the teacher immediately.

_____ Talk to the victim.

_____ Use CPR if necessary.

_____ Wait until help arrives before taking any action.

  Have your instructor check your answers and explain any answers that need correction.

Instructor's Score or Approval _____

Job Sheet 5-8

Name _____ Date _____

Identifying the Procedure to Use in Case of Fire

Objective

Upon completing this activity you will have demonstrated the ability to identify the procedure to use in case of a fire in the shop.

Tools and Materials

    Job sheet 5-8

    Pen or pencil

Procedure

Write Yes or No next to these actions to take in case of a fire.

_____ Avoid causing others to panic.

_____ Call everyone over to the fire scene.

_____ Clear the area.

_____ Call the fire department.

_____ Carry tools out of the shop.

_____ Notify the teacher.

_____ Scream at everyone.

_____ Set off the fire alarm.

_____ Try to keep others calm.

_____ Use any fire extinguisher at hand.

_____ Use the appropriate fire extinguisher if it can extinguish the fire.

Have your instructor check your answers and explain any answers that need correction.

Instructor's Score or Approval _____

Job Sheet 5-9

Name _____ Date _____

Identifying Requirements for the Use of the Slow Moving Vehicle (SMV) Emblem

Objective

Upon completing this activity you will have demonstrated your ability to identify the requirements for the SMV emblem's use.

Tools and Materials

    Job sheet 5-9

    Pen or pencil

Procedure

    Check the items that apply to the slow moving vehicle emblem requirements.

    _____ Any vehicle that travels less than 25 miles per hour on public highways.

    _____ Reflective orange triangle with a red border.

    _____ Is displayed at the front of the vehicle.

    _____ Is displayed on the rear of the vehicle.

    _____ It does not apply to towed wagons.

    _____ It applies to any farm machine towed at 25 mph or less.

    _____ Faded or damaged emblems should be replaced.

Instructor's Score or Approval _____

Unit 6 Shop Cleanup and Organization

When many people use the same facility and tools it is obvious that some order is essential so each person has the ability to find the tools that are necessary for their activities. It is also apparent that safety is important and a clean, well-organized work place promotes safety.

Team work and cooperation are needed in shop cleanup. Each student needs to be responsible for the work area used by that student as well as participating in the general clean up of the whole area.

Read Unit 6 and refer to it as needed while completing the following activities.

Shop Activity 6-1

Identifying the Shop Organization Aids

Shop Activity 6-2

Identifying a Suitable Shop Cleanup Plan

Job Sheet 6-1

Name _____ Date _____

Identifying the Shop Organization Aids

Objective

Upon completing this activity you will have demonstrated the ability to identify the shop organization aids.

Tools and Materials

    Job sheet 6-1

    Pen or pencil

Procedure

    Check the shop organization aids present in the shop.

____ Broom racks

____ Dust collection system

____ Fenced outside storage area

____ Locked tool panels

____ Lumber storage racks

____ Metal cans for rags

____ Metal storage for paints and materials

____ Metal storage racks

____ Paint room or booth

____ Parts cleaner

____ Project storage area

____ Scrap wood container

____ Shop vacuum

____ Shop cleanup chart or sheet

____ Tool storage room

____ Trash containers

____ Waste metal storage bin or containers

Instructor's Score or Approval _____

Job Sheet 6-2

Name _____ Date _____

Identifying a Suitable Shop Cleanup Plan

Objective

When you have completed this activity you will have demonstrated the ability to identify a suitable cleanup plan.

Tools and Materials

Job sheet 6-2

Pen or pencil

Procedure

Check the elements present in your shop cleanup plan.

____ All students participate.

____ All students share responsibility.

____ All tools are put away and accounted for.

____ Benches are cleaned.

____ Cabinets and storage areas are locked.

____ Cleaning items are properly stored.

____ Cleanup starts on signal.

____ Floors are cleaned.

____ Jobs are checked for completion.

____ Job responsibilities rotate.

____ Machines are vacuumed or cleaned.

____ Projects are properly stored.

____ Students are evaluated.

____ Students wait for instructor's dismissal.

____ Sinks and restrooms are clean and orderly.

____ Supplies are properly stored.

Instructor's Score or Approval _____

SECTION 3 HAND WOODWORKING AND METALWORKING

Unit 7 Hand Tools, Fasteners, and Hardware

A basic knowledge of hand tools and their use is essential to most mechanical jobs as some portion will often require their use. You should be able to identify the common shop tools and understand their uses. You should also learn to use hand tools before moving on to power tool operation.

Class Activity 7-1

Identifying the Classes of Tools

Class Activity 7-2

Classifying Tools

Class Activity 7-3

Identifying Common Fasteners

Class Activity 7-4

Identifying Common Hardware

Job Sheet 7-1

Name _____ Date _____

Identifying the Classes of Tools

Objective

Upon the completion of this activity you will be able to determine the classification of tools by their use.

Tools and Materials

Job sheet 7-1

Pen or pencil

Procedure

Match the uses with the classifications below.

___ 1. Layout tools (L)   a. Tools used to move another tool using force.

___ 2. Cutting tools (C)   b. Tools and extensions used to turn nuts or bolts and screws.

___ 3. Boring tools (B)   c. Tools that do not fit the other categories.

___ 4. Driving tools (Dr)   d. Tools used to measure or mark materials.

___ 5. Holding tools (H)   e. Tools used to loosen or remove earth.

___ 6. Turning tools (T)   f. Tools used to make holes or change their shapes.

___ 7. Digging tools (D)   g. Tools used to hold materials for the use of other tools.

___ 8. Other tools (O)   h. Tools used to cut or remove material.

Instructor's Score or Approval _____

Job Sheet 7-2

Name _____ Date _____

Classifying Tools

Objective

Upon completing this activity you will have demonstrated the ability to identify the classification of a tool.

Tools and Materials

Job sheet 7-2

Pen or pencil

Procedure

Match the tools with their proper classifications using the letter B for boring tools, C for cutting tools, D for digging tools, Dr for driving tools, H for holding tools, L for layout tools, O for other tools, and T for turning tools.

___ 1. Adjustable wrench
___ 2. Axes
___ 3. Backsaw
___ 4. Ball peen hammer
___ 5. Bar clamp
___ 6. Bench vise
___ 7. Bevel
___ 8. Blacksmith's hammer
___ 9. Block clamp
___ 10. Bolt cutters
___ 11. Bow saw
___ 12. Brick set
___ 13. Brick trowel
___ 14. Caliper rule
___ 15. Cape chisel
___ 16. Carpenter's square
___ 17. C Clamp
___ 18. Center punch
___ 19. Chalk line
___ 20. Cold chisel
___ 21. Combination square
___ 22. Compass saw
___ 23. Compound snips
___ 24. Concrete edger
___ 25. Concrete groover
___ 26. Coping saw
___ 27. Cordless power screwdriver
___ 28. Corner clamp
___ 29. Cultivator
___ 30. Countersink
___ 31. Curved-claw hammer
___ 32. Cutter mattock
___ 33. Deep socket
___ 34. Diagonal cutters
___ 35. Diamond-point chisel
___ 36. Die
___ 37. Diestock
___ 38. Dividers
___ 39. Drive extension
___ 40. Fencing tool
___ 41. Finishing trowel
___ 42. Flat file
___ 43. Flex handle
___ 44. Flex socket
___ 45. Folding wood rule
___ 46. Garden hoe
___ 47. Garden rake
___ 48. Glass cutter
___ 49. Grease gun
___ 50. Groove joint pliers
___ 51. Hacksaw
___ 52. Handsaw, crosscut
___ 53. Handsaw, rip
___ 54. Hand trowel
___ 55. Half-round file
___ 56. Hatchet

___	57.	Hex keys	___	58.	Hole saw
___	59.	Inside caliper	___	60.	Leaf gauge
___	61.	Level	___	62.	Lever wrench pliers
___	63.	Line level	___	64.	Long handled shovel
___	65.	Long handled spade	___	66.	Long nose pliers
___	67.	Machinist's vise	___	68.	Marking gauge
___	69.	Masonry drill	___	70.	Mason's hammer
___	71.	Mini hacksaw	___	72.	Nippers
___	73.	Offset box wrench	___	74.	Open end wrench
___	75.	Outside caliper	___	76.	Pipe cutters
___	77.	Pipe wrench	___	78.	Phillips screwdriver bit
___	79.	Phillips screwdriver	___	80.	Pick mattock
___	81.	Pin punch	___	82.	Pipe vise
___	83.	Plastic mallet	___	84.	Plug cutter
___	85.	Pointing trowel	___	86.	Post hole digger
___	87.	Power bit	___	88.	Power drill
___	89.	Prick punch	___	90.	Pruning saw
___	91.	Pruning shears	___	92.	Push drill
___	93.	Pry bar	___	94.	Ratchet offset screwdriver
___	95.	Ratchet handle	___	96.	Ripping chisel
___	97.	Ripping claw hammer	___	98.	Round file
___	99.	Rubber mallet	___	100.	Screwmate
___	101.	Scratch awl	___	102.	Scraper
___	103.	Sledge hammer	___	104.	Sliding T handle
___	105.	Slip joint pliers	___	106.	Socket adapter
___	107.	Spading fork	___	108.	Speed handle
___	109.	Splitting wedge	___	110.	Spring clamp
___	111.	Standard screwdriver bit	___	112.	Standard screwdriver
___	113.	Starter punch	___	114.	Steel tape
___	115.	Surform	___	116.	Tap
___	117.	Tap wrench	___	118.	Tinsnips
___	119.	Torque wrench	___	120.	Triangular file
___	121.	Try square	___	122.	Turf edger
___	123.	Twist drill	___	124.	Utility knife
___	125.	Universal joint	___	126.	Wire gauge
___	127.	Wood chisel	___	128.	Wrecking bar

Instructor's Score or Approval _____

Job Sheet 7-3

Name _____ Date _____

Identifying Common Fasteners

Objective

Upon the completion of this activity you will have demonstrated the ability to identify the common fasteners used in agricultural mechanics.

Tools and Materials

Job sheet 7-3
Pen or pencil

Procedure

Using the checklists, identify each of the common fasteners.

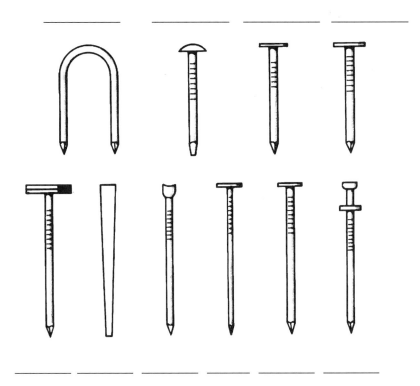

Figure 7-1. Types of Nails

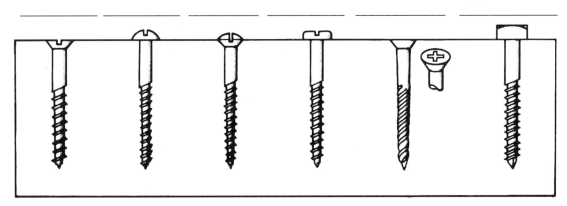

Figure 7-2. Screw Classifications

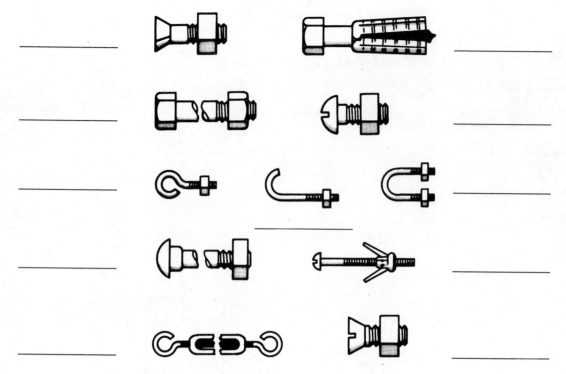

Figure 7-3. Common Bolts

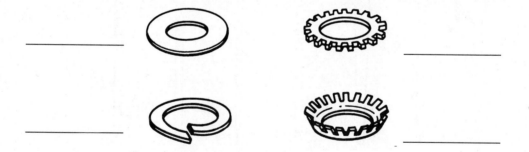

Figure 7-4. Types of Washers

Instructor's Score or Approval _____

Job Sheet 7-4

Name _____ Date _____

Identifying Common Hardware

Objective

Upon the completion of this activity you will have demonstrated the ability to identify the common hardware items.

Tools and Materials

    Job sheet 7-4

    Pen or pencil

Procedure

Using the checklist, identify the common hardware items.

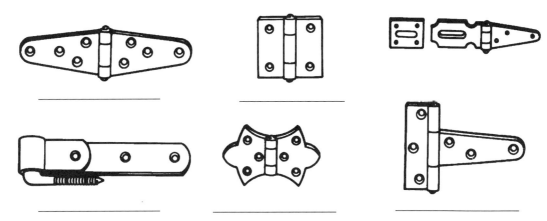

Figure 7-5. Types of Hinges

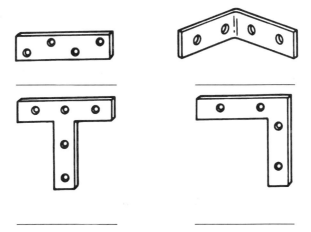

Figure 7-6. Flush Plates

Instructor's Score or Approval _____

Unit 8 Layout Tools and Procedures

Layout tools are important in designing and planning projects for shop construction. This unit will help you develop the ability to properly use layout tools.

Layout tools are made of metal, plastic, wood, and sometimes cloth. Metal provides rigidity and durability, plastic provides economical cost and light weight but is less durable and easily damaged by heat or chemicals. Wood is usually used for handles, but is not usually used for layout tools because of the ease of breakage or damage. Cloth is sometimes used in longer measuring tapes to lessen weight and bulk.

Class Activity 8-1

Identifying Inch and Metric Ruler Markings

Class Activity 8-2

Identifying Layout Tools

Class Activity 8-3

Identifying the Uses of Layout Tools

Shop Activity 8-4

Using a Square

Shop Activity 8-5

Squaring a Large Area

Shop Activity 8-6

Leveling an Object

Shop Activity 8-7

Creating a Pattern

Job Sheet 8-1

Name _____ Date _____

Identifying Inch and Metric Ruler Markings

Objective

Upon completing this activity you will have demonstrated the ability to identify the markings on inch and metric rulers.

Tools and Materials

Job sheet 8-1

Pen or pencil

Procedure

Fill in the blanks on the accompanying figures with the correct inch or metric readings.

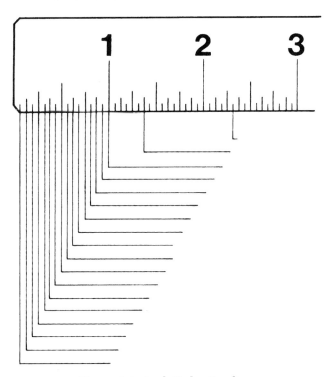

Figure 8-1. Inch Ruler Readings

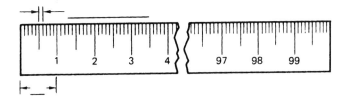

Figure 8-2. Metric Ruler Readings

Instructor's Score or Approval _____

Job Sheet 8-2

Name _____ Date _____

Identifying Layout Tools

Objective

Upon completing this activity you will have demonstrated the ability to identify layout tools.

Tools and Materials

Job sheet 8-2

Pen or pencil

Procedure

Using the accompanying figures, write in the correct names for the layout tools pictured.

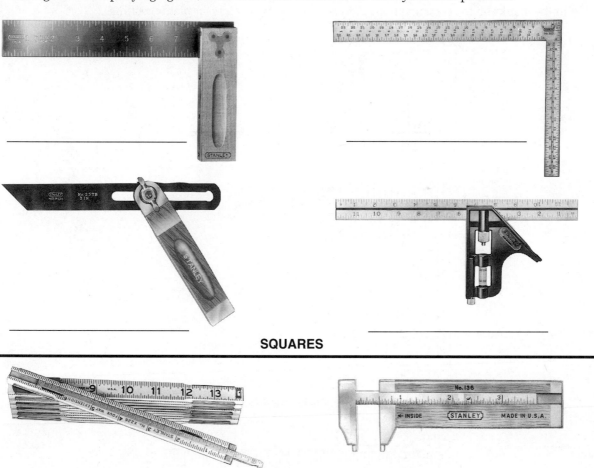

Figure 8-3. Layout Tools

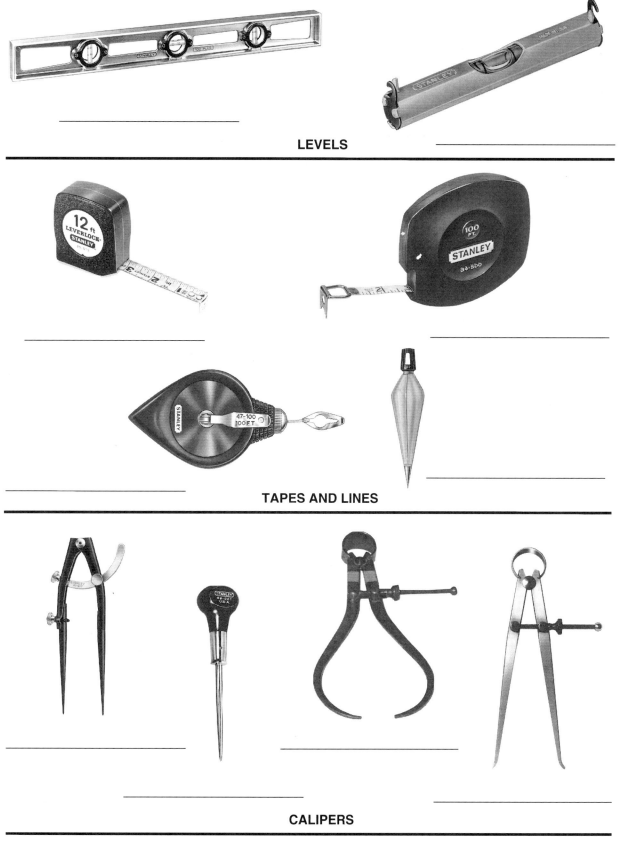

LEVELS

TAPES AND LINES

CALIPERS

Figure 8-3. Layout Tools (Continued)

Instructor's Score or Approval _____

> **Job Sheet 8-3**

Name _____ Date _____

Identifying the Uses of Layout Tools

Objective

Upon completion of this activity you will have demonstrated the ability to identify the uses of layout tools.

Tools and Materials

Job sheet 8-3

Pen or pencil

Procedure

Match the layout tool in the first column with the uses in the second column.

Tool *Layout Use*

___ 1. Bench rule or scale a. long straight lines

___ 2. CAD Program b. curved lines or circles

___ 3. Caliper rule c. measuring up to 2 to 8 feet

___ 4. Carpenter's square d. measuring thickness or gaps

___ 5. Chalk line e. measuring inside dimensions

___ 6. Combination square f. establishing equal heights

___ 7. Dividers and scribers g. guide for laying blocks

___ 8. Folding rule h. very accurate small measuring

___ 9. Gauges i. measuring outside diameters

___ 10. Inside caliper j. computer assisted design

___ 11. Level or line level k. transfer a vertical point

___ 12. Line l. construction framing

___ 13. Outside caliper m. rigid measuring tool

___ 14. Pattern n. transferring or marking angles

___ 15. Plumb bob o. copying objects

___ 16. Try square p. testing accuracy of cuts

___ 17. Sliding T bevel q. flexible measuring rule

___ 18. Tape rule r. versatile measuring or testing tool

Instructor's Score or Approval _____

Job Sheet 8-4

Name _____ Date _____

Using a Square

Objective

Upon completing this activity you will have demonstrated the ability to use a square to mark a board and to check the accuracy of a saw cut.

Tools and Materials

    Job sheet 8-4

    Combination or try square

    Handsaw

    Sawhorse

    Board

    Pencil

Procedure

1. Place the board on the sawhorse.
2. Place the square on the board 1/2 inch from the end.
3. Mark along the blade of the square.
4. Turn the board on edge and place the square at the start of the first mark.
5. With the board and marks extending over the end of the sawhorse, make the saw cut.
6. Use the square to check the accuracy of the cut.

Instructor's Score or Approval _____

Job Sheet 8-5

Name _____ Date _____

Squaring a Large Area

Objective

Upon the completion of this activity you will have demonstrated the ability to square a large area.

Tools and Materials

Job sheet 8-5

Carpenter's square

50' tape

String

Stakes

Procedure

1. Set a stake to establish a first corner.
2. Place the heel of the square against the stake.
3. Attach a string and stretch it along one leg of the square and stake the other end at 20 feet.
4. Attach a second string and stretch it along the other leg at a right angle and stake it at 15 feet.
5. Measure from the corner six feet and tie a piece of string at that point.
6. Measure from the corner eight feet at a right angle and tie a piece of string at that point.
7. Measure between the two pieces of string.
8. Move one string in or out until the distance is exactly 10 feet or within 1/16 of an inch.
9. Place the heel of the square at the third stake with one leg against the second string and the other pointing in the same direction as the second stake.
10. Stretch a string along the leg of the square and stake it at 20 feet.
11. Repeat steps 5-8.
12. Measure 15 feet to the second stake.
13. Measure the diagonals and adjust corners until they are equal.

Instructor's Score or Approval _____

Job Sheet 8-6

Name _____ Date _____

Leveling an Object

Objective

Upon completion of this activity you will have demonstrated the ability to level an object.

Tools and Materials

Job sheet 8-6

Level

Sawhorse

Shims

An uneven surface

Procedure

1. Set the sawhorse on the uneven surface.
2. Set the level on the sawhorse.
3. Place shims under the legs furthermost from the bubble in the level to center the bubble.
4. Repeat placing shims until the bubble is centered in the level.
5. Replace the sawhorse and level.

Instructor's Score or Approval _____

Job Sheet 8-7

Name _____ Date _____

Creating a Pattern

Objective

Upon completing this activity you will have demonstrated the ability to create a pattern.

Tools and Materials

　Job sheet 8-7

　Picture of step 2, page 95 in the text

　8″ x 10″ tag board

　Ruler

　Pencil

Procedure

1. Fold the tag board in half lengthwise.
2. Make a grid with 1″ spacing.
3. Draw the tree on the enlarged grid.
4. Cut out the tree.
5. Unfold the tree.

Instructor's Score or Approval _____

Unit 9 Selecting, Cutting, and Shaping Wood

Lumber is the most common building material today because of its low cost and workability. You should become familiar with its characteristics and its shaping.

Although power saws have usually replaced hand saw use, the principles and practices used in these exercises will be useful as you later learn the use of power tools.

Shop Activity 9-1

Identifying Common Lumber Species

Shop Activity 9-2

Identifying Nominal and Actual Lumber Dimensions

Shop Activity 9-3

Identifying Handsaws

Shop Activity 9-4

Using a Crosscut Handsaw

Shop Activity 9-5

Using a Rip Saw

Shop Activity 9-6

Using a Surform to Smooth an Edge

Shop Activity 9-7

Making a Dado

> **Job Sheet 9-1**

Name _____ Date _____

Identifying Common Lumber Species

Objective

Upon completing this activity you will have demonstrated the ability to identify common species of lumber.

Tools and Materials

    Lumber samples of common species

    Job sheet 9-1

    Pencil or pen

Procedure

Take the samples of the common species provided by your instructor and complete the following information table by observation. Use as many species as your instructor has available. Study the samples until you can correctly identify other samples supplied without identifications. Complete the remainder of the table by using the textbook.

Wood	Hard	Medium	Soft	Color
Birch	___	___	___	_____
Cedar, Red	___	___	___	_____
Cherry	___	___	___	_____
Cypress	___	___	___	_____
Fir	___	___	___	_____
Hemlock	___	___	___	_____
Maple	___	___	___	_____
Mahogany	___	___	___	_____
Oak	___	___	___	_____
Pine, Yellow	___	___	___	_____
Pine, White	___	___	___	_____
Redwood	___	___	___	_____
Walnut, Black	___	___	___	_____
Willow, Black	___	___	___	_____
_____	___	___	___	_____
_____	___	___	___	_____
_____	___	___	___	_____

Instructor's Score or Approval _____

Job Sheet 9-4

Name _____ Date _____

Using a Crosscut Handsaw

Objective

Upon the completion of this activity you will have demonstrated the ability to properly use a crosscut saw.

Tools and Materials

Job sheet 9-4

1" x 6" x 18"

Crosscut handsaw

Pencil

Try square

Sawhorse

Procedure

1. Place the 1" x 6" on the sawhorse.
2. Place the try square against the edge of the board and mark a line 1/2" from the end.
3. Turn the 1" x 6" on edge and mark a line with the square touching the start of the first line.
4. Extend the board over the end of the sawhorse so the line is beyond the sawhorse.
5. Hold the board on the sawhorse with the left knee.*
6. Hold the saw with the index finger along the handle.
7. Align the right shoulder, knee, elbow, and eye with the line.**
8. Place the thumbnail on the mark to aid the start of the cut.
9. Place the saw on the mark at a 45 degree angle so you see the back of the saw and the line aligned.
10. Start the cut with a backstroke.
11. Saw the board with long even strokes with no pressure on the last stroke.
12. Check the board with the try square for accuracy and squareness.
13. Repeat the steps until you produce square cuts.
14. Clean the area and put away the tools.

    *Use the right knee if you are left-handed.

    **Use the left shoulder, knee, elbow, and eye if you are left-handed.

Instructor's Score or Approval _____

	Job Sheet 9-5

Name _____ Date _____

Using a Ripsaw

Objective

Upon completing this activity you will have demonstrated the ability to use a ripsaw.

Tools and Materials

Job sheet 9-5

The board used in activity 9-4

Ripsaw

Sawhorse

Combination square

Pencil

Procedure

1. Place the board on the sawhorse.
2. Set the combination square at 4 1/8".
3. Place the pencil at the end of the square and draw a line the length of the board.
4. Rip the board using the technique from activity 9-4.
5. Check the board for a square cut.
6. Clean the area and put the tools away.

Instructor's Score or Approval _____

Job Sheet 9-6

Name _____ Date _____

Using a Surform to Smooth an Edge

Objective

Upon the completion of this activity you will have demonstrated the ability to use a surform to smooth an edge.

Tools and Materials

Job sheet 9-6

The board used in previous activities

Surform

Combination square

Bench vise

Pencil

Procedure

1. Set the combination square at 4".
2. Mark the length of the board on the side of the rip cut.
3. Mark the length of the board on the other side.
4. Place the board in the vise.
5. Push the surform at an angle with the grain with long smooth strokes.
6. Check the edge for squareness.
7. Remove any additional wood until the edge is straight, smooth, and square.
8. Clean the area and put away the tools.

Instructor's Score or Approval _____

Job Sheet 9-7

Name _____ Date _____

Making a Dado

Objective

Upon the completion of this activity you will have demonstrated the ability to make a dado.

Tools and Materials

Job sheet 9-7

Crosscut saw

Two pieces of 1" x 6" lumber

1/2" wood chisel

Combination square

Bench vise

Pencil

Procedure

1. Draw two parallel lines across one of the boards exactly 3/4" apart with the square.
2. Mark across both edges, square with each of the two marks.
3. Set the combination square at 3/8" and mark the depth of cut on each edge of the board.
4. Place the board in the bench vise.
5. Saw a kerf on each line to the depth of cut.
6. Saw additional kerfs to the depth of cut.
7. Remove the wood between the kerfs with the wood chisel.
8. Fit the second board into the dado.
9. Remove more wood if needed to perfect the fit.
10. Clean the area and put away the tools.

Instructor's Score or Approval _____

Unit 10 Fastening Wood

Wood is very important in agriculture for construction of buildings, fences, parts of tools and machinery. The ability to fasten wooden parts with strong joints is a valuable skill. This unit is concerned with the types of joints and fasteners that keep the joints rigid and secure.

Class Activity 10-1

Identifying Common Wood Joints

Class Activity 10-2

Understanding Nailing Techniques

Shop Activity 10-3

Fastening Wood with Nails

Shop Activity 10-4

Fastening Wood with Screws

Shop Activity 10-5

Fastening Wood with Bolts

Shop Activity 10-6

Fastening Wood with Glue and Screws

Shop Activity 10-7

Fastening Wood with Dowels and Glue

Job Sheet 10-1

Name _____ Date _____

Identifying Common Wood Joints

Objective

Upon completing this activity you will have demonstrated the ability to identify the common wood joints by description or pictures.

Tool and Materials

Job sheet 10-1

Pen or pencil

Procedure

Match the description of the joint with the name of the joint, then label the pictures of the joints correctly.

___ 1. Butt joint   a. A joint made by interlocking parts of two pieces.

___ 2. Lap joint   b. A joint made by placing two pieces end to end or side to side.

___ 3. Dado and rabbet joint   c. A joint made by cutting the ends of two pieces at 45 degrees then joining at 90 degrees.

___ 4. Miter joint   d. A joint formed by fastening one piece face to face with another.

___ 5. Dovetail joint   e. A joint made by making a rectangular groove in a board or at the end to receive the end of the second piece.

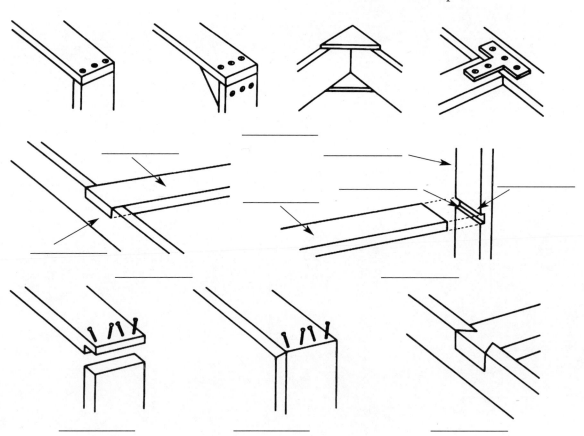

Figure 10-1. Common Wood Joints

Instructor's Score or Approval _____

Job Sheet 10-2

Name _____ Date _____

Understanding Nailing Techniques

Objective

Upon the completion of this activity you will have demonstrated an understanding of nailing techniques.

Tools and Materials

Job sheet 10-2

Pen or pencil

Procedure

Complete the following work sheet.

1. Why is a block of wood recommended when pulling nails?
   _____
   _____

2. Why is a nail set used in finish work?
   _____
   _____

3. What is toe nailing and when is it used?
   _____
   _____
   _____

4. What is end nailing and when is it used?
   _____
   _____
   _____

5. What is clinching and why is it used?
   _____
   _____

6. What are staples and how are they used?
   _____
   _____
   _____
   _____
   _____

Instructor's Score or Approval _____

Job Sheet 10-3

Name _____ Date _____

Fastening Wood with Nails

Objective

Upon the completion of this activity you will have demonstrated the ability to fasten wood with nails.

Tools and Materials

Job sheet 10-3

Two pieces of 2″ x 4″ wood

Curved claw hammer

8d common nails

Procedure

1. Place one piece of wood across the second piece so the two ends are even with the sides.
2. Select a point for the first nail by staying the width of the board from the edge of the board.
3. Start a nail by tapping it lightly at a 90 degree angle to the board.
4. Drive the nail with the hammer held at the end of the handle.
5. Use the wrist and arm to deliver firm blows.
6. Have the face of the hammer parallel to the board when the blow is struck.
7. Drive the nails until the head is flush with the wood and the two boards are securely fastened.
8. Remove the nails and put away the tools.

Instructor's Score or Approval _____

Job Sheet 10-4

Name _____ Date _____

Fastening Wood with Screws

Objective

Upon completing this activity you will have demonstrated the ability to fasten wood with screws.

Tools and Materials

Job sheet 10-4

A piece of 1″ x 4″

A piece of 2″ x 4″

Standard screwdriver

Four No. 8 x 2″ flat head screws

Countersink

Shank hole drill

Pilot hole drill

Power drill

Procedure

1. Place the 1″ x 4″ across the end of the 2″ x 4″ with the ends even with the edges.
2. Mark the four spots for the screws.
3. Select the proper shank hole drill (see text Figure 10-8).
4. Install the drill in the power drill.
5. Drill the shank holes.
6. Mark the location of the pilot holes.
7. Install the pilot hole drill.
8. Drill the pilot holes 1″ to 1 1/4″ deep.
9. Install the countersink in the drill.
10. Countersink the shank holes so the screws will be flush when tightened.
11. Install the screws and turn them until the heads are flush.
12. Clean the area and put away the tools.

Instructor's Score or Approval _____

Job Sheet 10-5

Name _____ Date _____

Fastening Wood with Bolts

Objective

Upon completing this activity you will have demonstrated the ability to fasten wood with bolts.

Tools and Materials

Job sheet 10-5

Two 5/16" x 3 1/2" carriage bolts

Two 2" x 4" boards

Power drill

A 5/16 drill bit

Two 5/16" washers and nuts

Bench vise

1/2" wrench

Procedure

1. Mark the locations for the bolts.
2. Install the board in the vise to drill the holes.
3. Install the auger bit in the brace.
4. Drill the first 2" x 4" remembering to complete the hole from the back side.
5. Mark and drill the second 2" x 4".
6. Install the two carriage bolts.
7. Tighten the bolts snugly without crushing the wood fibers and drawing the head below the surface.
8. Clean the area and put away the tools.

Instructor's Score or Approval _____

Job Sheet 10-6

Name _____ Date _____

Fastening Wood with Glue and Screws

Objective

Upon completing this activity you will have demonstrated the ability to fasten wood with glue and screws.

Tools and Materials

    Job sheet 10-6

    Two pieces of wood

    Wood screws

    Aliphatic resin glue

    Putty knife

    Drill

    Pencil

    Try square

Procedure

1. Cut the two pieces of wood so surfaces match perfectly.
2. Modify the wood if needed to secure a complete match.
3. Drill the screw holes.
4. Apply a small bead of glue on each surface.
5. Spread the glue evenly.
6. Install the screws.
7. Remove excess glue or glue runs with the putty knife.
8. Clean the area and put away the materials.

Instructor's Score or Approval _____

Job Sheet 10-7

Name _____ Date _____

Fastening Wood with Dowels and Glue

Objective

Upon completion of this activity you will have demonstrated the ability to fasten wood with dowels and glue.

Tools and Materials

Job sheet 10-7

Two pieces of wood with straight edges

Two bar clamps

3/8" dowel stock

Backsaw

Sandpaper

Aliphatic resin glue

Combination square

3/8" auger bit

Bit brace

Bench vise

Putty knife

Pencil

Procedure

1. Prepare perfectly fitting edges.
2. Place the two faces of the boards together in the vise with the ends flush.
3. Mark across the two pieces approximately 1/3 the distance from each end.
4. Set the combination square at 3/8".
5. Mark each line at 3/8".
6. Drill 1 1/8" deep hole at each center mark.
7. Cut two 2" dowels.
8. Taper the ends of the dowels with sandpaper.
9. Test fit the pieces.
10. Apply a thin bead of glue to each edge and into the holes.
11. Assemble the pieces and clamp.
12. Remove any glue runs.
13. Clean the area and put away the tools.
14. The next day remove the clamps and inspect the joint.
15. Put away the clamps.

Instructor's Score or Approval _____

Unit 11 Finishing Wood

Finishing materials protect wood from weathering and other damage. Paints, stains, varnish, and wax are among the common wood finishes. Proper finishing requires the preparation of the wood to receive the finish. In this unit you will learn about finishing materials and methods.

Shop Activity 11-1

Removing Dents from Wood

Class Activity 11-2

Identifying Common Sanding Materials

Shop Activity 11-3

Sanding Wood

Shop Activity 11-4

Filling Holes in Wood with Putty or Glazing Compound

Shop Activity 11-5

Filling Holes in Wood with Plastic Wood

Shop Activity 11-6

Applying a Stain to Wood

Shop Activity 11-7

Applying a Wood Sealer

Shop Activity 11-8

Applying a Finish Material to Wood

Shop Activity 11-9

Cleaning and Caring for Brushes

Job Sheet 11-1

Name _____ Date _____

Removing Dents from Wood

Objective

Upon completing this activity you will have demonstrated the ability to remove dents from wood.

Tools and Materials

Job sheet 11-1

Dented wood

Soldering copper

Damp soft cloth

Procedure

1. Heat the soldering copper to 400 degrees.
2. Place the damp cloth over the dent.
3. Apply the heat for 5 seconds until steam arises.
4. Remove the copper and cloth.
5. Examine the dent and if still evident repeat steps 1–4.
6. Sand when the wood is dry.

Instructor's Score or Approval _____

Job Sheet 11-2

Name _____ Date _____

Identifying Common Sanding Materials

Objective

Upon completing this activity you will have demonstrated the ability to identify the common sanding materials.

Tools and Materials

Job sheet 11-2

Pen or pencil

Procedure

Match the materials in column I with the descriptions of their use in column II.

	Column I		Column II
____ 1.	Aluminum oxide	a.	when clogging is a problem
____ 2.	Emery	b.	hand sanding
____ 3.	Flint	c.	speed and durability
____ 4.	Garnet	d.	sanding metal
____ 5.	Silicone carbide	e.	wet sanding
____ 6.	Steel wool	f.	smoothing and polishing

Instructor's Score or Approval _____

Job Sheet 11-3

Name _____ Date _____

Sanding Wood

Objective

Upon completing this activity you will have demonstrated the ability to sand wood.

Tools and Materials

Job sheet 11-3

Medium, fine, and very fine sandpaper

Sanding block

Wood to be sanded

Bench vise

Procedure

1. Wrap the sanding block with medium sandpaper.
2. Place the wood in the bench vise so the face is above the jaws.
3. Sand with the grain, moving over the whole surface of the board.
4. Sand each surface of the board.
5. Change to fine sandpaper and repeat.
6. Change to very fine sandpaper and complete the sanding.
7. Inspect the surfaces and resand as needed.
8. Clean the area and put away the materials.

Instructor's Score or Approval _____

Job Sheet 11-4

Name _____ Date _____

Filling Holes in Wood with Putty or Glazing Compound

Objective

Upon completing this activity you will have demonstrated the ability to fill holes in wood with putty or glazing compound.

Tools and Materials

Job sheet 11-4

Wood with holes to be filled

Putty or glazing compound

Putty knife

Standard screwdriver

Tinting material

Procedure

1. Seal the wood.
2. Work a small amount of material in your hand until it is soft.
3. Tint the material to the desired shade.
4. With the screwdriver tip force a bit of material into the hole.
5. Smooth with the putty knife.
6. Clean the area and put away the tools and materials.
7. Let it set overnight.
8. Apply the desired finish.

Instructor's Score or Approval _____

Job Sheet 11-5

Name _____ Date _____

Filling Holes in Wood with Plastic Wood

Objective

Upon completing this activity you will have demonstrated the ability to fill holes in wood with plastic wood.

Tools and Materials

    Job sheet 11-5

    Wood with holes to be filled

    Plastic wood of the desired color

    Putty knife

    Standard screwdriver

Procedure

1. Put a small amount on the tip of the screwdriver and close the tube or can.
2. Quickly force the material into the hole.
3. Smooth the material with the putty knife.
4. Add more material if needed.
5. Recheck to be sure the can is tightly closed.
6. Scrape off any material around the hole.
7. Wait 3 to 5 minutes for the material to dry.
8. Sand with fine sandpaper.
9. Apply the desired finish.
10. Clean the area and put away the materials.

Instructor's Score or Approval _____

Job Sheet 11-6

Name _____ Date _____

Applying a Stain to Wood

Objective

Upon completing this activity you will have demonstrated the ability to apply wood stains.

Tools and Materials

Job sheet 11-6

Wood to be stained

Stain of desired color

Clean soft rags

Brush

Procedure

1. Finish sanding the wood.
2. Stir the stain thoroughly.
3. Apply an even coat of stain with the brush.
4. Allow stain to stand 5 minutes.
5. Wipe the excess stain off with the soft rags.
6. Wipe and polish lightly until no stain comes off.
7. Clean the brush and put away the materials.
8. Allow 24 hours for drying.
9. Apply the desired finish.

Instructor's Score or Approval _____

Job Sheet 11-7

Name _____ Date _____

Applying a Wood Sealer

Objective

Upon the completion of this activity you will have demonstrated the ability to apply a wood sealer.

Tools and Materials

Job sheet 11-7

Wood to be sealed

Wood sealer

Brush

Tack rag (a rag dampened with a solvent)

Cleaning solvent

Procedure

1. Finish sanding the wood.
2. Use a tack rag to wipe the wood.
3. Read and follow the directions for preparing the sealing product.
4. Apply an even coat with a brush.
5. Clean the brush.
6. Clean the area and put away the materials.

Instructor's Score or Approval _____

Job Sheet 11-8

Name _____ Date _____

Applying a Finish Material to Wood

Objective

Upon completing this activity you will have demonstrated the ability to apply paint or varnish to wood.

Tools and Materials

Job sheet 11-8

Paint brush suitable to material to be used

Finish, paint, or varnish

Wood to be finished

Tack rag

Brush cleaning materials

Procedure

1. Sand the wood.
2. Read the label and follow the directions for preparing the finish for use.
3. Wipe the wood with a tack rag.
4. Dip the brush 1/2" and brush back and forth.
5. Alternate dipping and brushing to prevent drying in the brush.
6. Coat all surfaces to be protected with an even coat.
7. Do not bend the bristles of the brush or spread them excessively.
8. Clean the brush and put the materials away.

Instructor's Score or Approval _____

Job Sheet 11-9

Name _____ Date _____

Cleaning and Caring for Brushes

Objective

Upon completing this activity you will have demonstrated the ability to clean and care for brushes.

Tools and Materials

Job sheet 11-9

Solvents

Container just a little bigger than the brush

Used solvent container

Paint brush needing cleaning

Warm water

Soap

Paper towels

Procedure

1. Remove as much material as possible with rags or paper.
2. Pour 1/2" of solvent in the container.
3. Work the bristles up and down in the solvent.
4. Pour the solvent into the used solvent container.
5. Pour clean solvent and repeat steps 3 and 4.
6. Wipe the excess solvent from the brush with paper towels.
7. Work the soap into a lather and wash the brush until clean.
8. Rinse the soap from the brush.
9. Lay the brush on the lower left corner of a paper towel and wrap one or two turns, then fold the towel down over the brush and complete the wrap, then twist the towel around the handle.
10. Lay the brush aside to dry.

Instructor's Score or Approval _____

Unit 12 Identifying, Marking, Cutting, and Bending Metal

Since most agricultural machines are made of metal you need to learn the basics of metalworking.

Shop Activity 12-1

Identifying Metals

Shop Activity 12-2

Marking Metal

Shop Activity 12-3

Cutting Metal with a Hacksaw

Class Activity 12-4

Identifying Files

Class Activity 12-5

Identifying Metal Snips

Shop Activity 12-6

Cutting Metal with a Cold Chisel

Shop Activity 12-7

Shaping Metal with a File

Shop Activity 12-8

Bending Metal

Shop Activity 12-9

Rounding Metal

Shop Activity 12-10

Twisting Metal

Shop Activity 12-11

Bending Sheet Metal

Job Sheet 12-1

Name _____ Date _____

Identifying Metals

Objective

Upon the completion of this activity you will have demonstrated the ability to identify the distinguishing characteristics of metals.

Tools and Materials

Job sheet 12-1

Pen or pencil

Samples of metals

Procedure

Match the metals with the characteristics, have the answers corrected by your instructor, then identify the numbered samples.

____ 1. Aluminum
____ 2. Brass
____ 3. Bronze
____ 4. Cast iron
____ 5. Copper
____ 6. Galvanized steel
____ 7. Lead
____ 8. Mild steel
____ 9. Stainless steel
____ 10. Tin
____ 11. Tool steel
____ 12. Wrought iron

a. reddish brown, excellent conductor
b. soft, malleable, corrosion resistant
c. soft, malleable, corrosion resistant
d. very malleable, corrosion resistant, silver color
e. bright, hard, tough, corrosion resistant
f. soft, very heavy, bluish gray
g. silver white, good electric conductor, light, tough
h. zinc coated metal
i. high carbon, heat treatable, expensive
j. malleable, tough, rust resistant
k. malleable, ductile, tough
l. forms into any shape, brittle

Sample Number

1. _____   2. _____
3. _____   4. _____
5. _____   6. _____
7. _____   8. _____
9. _____   10. _____
11. _____   12. _____
13. _____   14. _____
15. _____

Study the samples you incorrectly identified until you can correctly identify them.

Instructor's Score or Approval _____

Job Sheet 12-2

Name _____ Date _____

Marking Metal

Objective

Upon completing this activity you will have demonstrated the ability to mark metals.

Tools and Materials

    Job sheet 12-2

    Soapstone

    Chalk

    Center punch

    Hammer

    Scriber

    Combination square

    Mild steel

Procedure

1. Place the combination square against a straight side of the mild steel.
2. Mark a line across the steel with the soapstone.
3. Using the scriber, mark a second line 1" from the first.
4. Place the tip of the center punch 1" from the edge of the metal.
5. Tap the punch lightly with the hammer.
6. Check to see if the prick mark is exactly 1" from the edge.
7. If the mark is incorrect, reset the punch.
8. With the punch in the correct spot, hit the punch with one solid blow.
9. With the chalk, make a third line on the metal.
10. Brush the metal.
11. Observe the best marks.
12. Clean the area and put away the materials.

Instructor's Score or Approval _____

Job Sheet 12-3

Name _____ Date _____

Cutting Metal with a Hacksaw

Objective

Upon completing this activity you will have demonstrated the ability to cut metal with a hacksaw.

Tools and Materials

Job sheet 12-3

Mild steel

Soapstone

Combination square

Machinist's vise

Hacksaw

Procedure

1. Place the combination square on a straight edge of the mild steel.
2. Mark a line across the steel with soapstone.
3. Place the mild steel in the vise with the line 1/2" from the jaws of the vise.
4. Select a hacksaw blade that will have 3 or more teeth on the metal as it cuts.
5. Cut the metal to the mark.
6. Apply pressure only on the forward stroke.
7. Check the squareness of your cut.
8. Save the metal for activity 12-7.
9. Clean the area and put away the materials.

Instructor's Score or Approval _____

Job Sheet 12-4

Name _____ Date _____

Identifying Files

Objective

Upon completing this activity you will have demonstrated the ability to identify files.

Tools and Materials

Job sheet 12-4

Pen or pencil

Procedure

Correctly label the files pictured below.

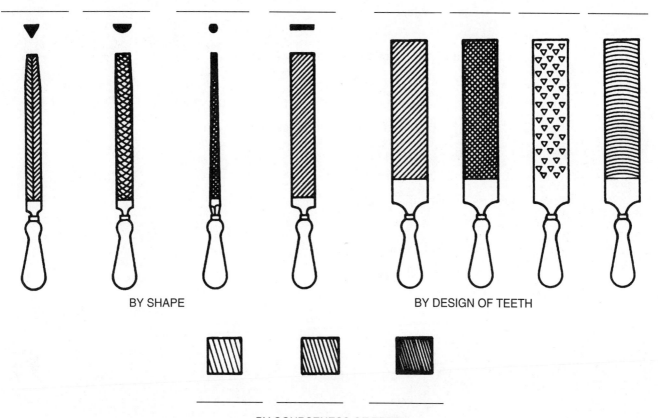

Figure 12-1. Identifying Files

Instructor's Score or Approval _____

Job Sheet 12-5

Name _____ Date _____

Identifying Metal Snips

Objective

Upon completing this activity you will have demonstrated the ability to identify metal snips.

Tools and Materials

    Job sheet 12-5

    Pen or pencil

Procedure

    Correctly label the snips pictured below.

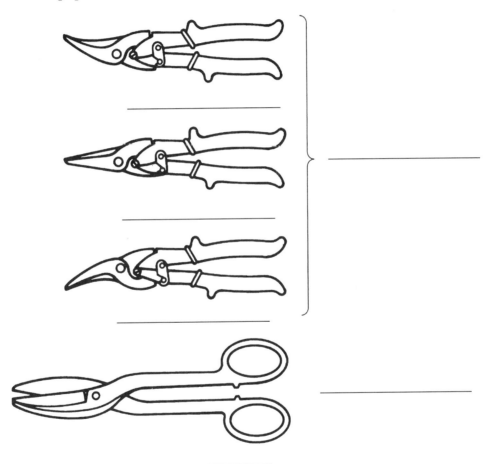

Figure 12-2. Identifying Metal Snips

Instructor's Score or Approval _____

Job Sheet 12-6

Name _____ Date _____

Cutting Metal with a Cold Chisel

Objective

Upon completing this activity you will have demonstrated the ability to cut metal with a cold chisel.

Tools and Materials

- Job sheet 12-6
- Mild steel
- Cold chisel
- Blacksmith's hammer
- Machinist's vise
- Combination square
- Soapstone
- Leather gloves

Procedure

1. Place the combination square against a straight edge of the steel.
2. Mark a line across the metal with the soapstone.
3. Place the metal in the vise with the mark even with the jaws of the vise.
4. Lay the heavy cold chisel against the corner of the iron at a 30 degree angle to the bench and 60 degrees to the metal.
5. Drive the chisel so it shears the metal.
6. Clean the area and put away the materials.

Instructor's Score or Approval _____

Job Sheet 12-7

Name _____ Date _____

Shaping Metal with a File

Objective

Upon completing this activity you will have demonstrated the ability to shape metal with a file.

Tools and Materials

- Job sheet 12-7
- Combination square
- Metal from activity 12-3
- Machinist's vise
- Scriber

Procedure

1. Check the hacksaw cut for squareness both crosswise and edgewise.
2. If the edge is approved for squareness, scribe a line 1/16" below the cut.
3. If the edge is not square, scribe a line that will take it to square.
4. File with the file tip in the fingers of one hand and the handle in the other hand.
5. File to the scribe mark applying no pressure on the backstroke.
6. Test for squareness frequently.
7. Clean the area and put away the materials.

Instructor's Score or Approval _____

Job Sheet 12-8

Name _____ Date _____

Bending Metal

Objective

Upon completing this activity you will have demonstrated the ability to bend metal.

Tools and Materials

    Job sheet 12-8

    Mild steel

    Combination square

    Soapstone

    Blacksmith's hammer

    Machinist's vise

Procedure

1. Mark the metal at the desired point of the bend.
2. Place the mark at the top of the vise jaws.
3. With one hand push the top of the metal in the direction of the bend while striking the metal at the vise with the hammer.
4. Drive the metal until the 90 degree angle is obtained.
5. Clean the area and put away the materials.

Instructor's Score or Approval _____

Job Sheet 12-9

Name _____ Date _____

Rounding Metal

Objective

Upon completing this activity you will have demonstrated the ability to round metal.

Tools and Materials

    Job sheet 12-9

    Mild steel 1" to 2" wide by 1/8" thick by 12" long

    Pipe or round stock

    Blacksmith's hammer

    Machinist's vise

    Leather gloves

Procedure

1. Place the end of the steel to be bent tightly between the round stock and the vise jaws.
2. With a hand and the hammer, bend the steel around the stock or pipe.
3. Release the vise and rotate the stock.
4. Repeat steps 2 and 3 until the desired roundness is attained.
5. Clean the area and put away the materials.

Instructor's Score or Approval _____

Job Sheet 12-10

Name _____ Date _____

Twisting Metal

Objective

Upon completing this activity you will have demonstrated the ability to twist metal.

Tools and Materials

Job sheet 12-10

Mild steel strap 3/4" to 1" wide by 1/8" thick by 12 or more inches long

12" adjustable wrench

Machinist's vise

Soapstone

Square

Procedure

1. Place the square where you want the twist to start.
2. Mark the start of the bend.
3. Measure 1 1/2 times the width of the metal and mark again.
4. Place the first mark at the top of the vise jaws.
5. Adjust the wrench to fit across the width of the steel.
6. Slide the wrench up to the second mark.
7. Hold the metal in one hand and twist with the other.
8. Stop at 90 degrees.
9. Clean the area and put away the materials.

Instructor's Score or Approval _____

Job Sheet 12-11

Name _____ Date _____

Bending Sheet Metal

Objective

Upon completing this activity you will have demonstrated the ability to bend sheet metal.

Tools and Materials

Job sheet 12-11

Sheet metal

Angle iron

Blacksmith's hammer

Clamps

Leather gloves

Procedure

1. Clamp the angle iron to the edge of the bench.
2. Place the sheet metal at the point where the bend is desired.
3. Start the bend with your hands.
4. Continue the bend tapping with the hammer until the bend is complete.
5. Clean the area and put away the materials.

Instructor's Score or Approval _____

Unit 13 Fastening Metal

This unit deals with three ways of fastening metals together. Metal fasteners are known as screws, bolts, and rivets. When metals are held together by melting another metal between them, it is known as soldering or brazing. When two metals are melted together, it is known as welding.

Shop Activity 13-1

Drilling Holes in Metal

Shop Activity 13-2

Tapping Threads in Metal

Shop Activity 13-3

Threading a Rod or Bolt

Shop Activity 13-4

Cutting and Threading Pipe

Shop Activity 13-5

Fastening Metal with Bolts and Nuts

Shop Activity 13-6

Fastening Metal with Cap Screws

Shop Activity 13-7

Fastening Metal with Rivets

Shop Activity 13-8

Fastening Metal with Sheet Metal Screws

Shop Activity 13-9

Tinning a Soldering Copper

Shop Activity 13-10

Soldering Sheet Metal

Shop Activity 13-11

Soldering Electrical Connections

Shop Activity 13-12

Soldering Copper Pipes

> **Job Sheet 13-1**

Name _____ Date _____

Drilling Holes in Metal

Objective

When you have completed this activity you will have demonstrated the ability to drill holes in metal.

Tools and Materials

  Job sheet 13-1
  Drill press
  High speed twist bit
  Face shield
  Leather gloves
  Protective clothing
  Drill press vise
  Metal to be drilled
  Combination square
  Center punch
  Ball peen hammer
  Clamp
  Cutting oil
  Metal file

Procedure

1. Obtain permission.
2. Put on the face shield.
3. Put on the leather gloves.
4. Put on the protective clothing.
5. Measure and mark the spot to be drilled.
6. Center punch the spot with a light tap.
7. Check the position and adjust if needed and give the center punch one sharp blow.
8. Set the proper drill press speed for the size of the bit.
9. If the hole is over 3/8" in diameter, use a 1/4" pilot hole.
10. Tighten the drill in the chuck using at least two holes.
11. Remove the chuck wrench.
12. Clamp the metal in the vise or hold it against the column or clamp it to the table.
13. Lower the quill and drill the hole with even pressure.
14. Use cutting oil to cool the bit.
15. Slacken the pressure as the drill breaks through.
16. Remove any burrs with the metal file.
17. Clean the area and put away the materials.

Instructor's Score or Approval _____

Job Sheet 13-2

Name _____ Date _____

Tapping Threads in Metal

Objective

When you have completed this activity you will have demonstrated the ability to thread a hole in metal.

Tools and Materials

Job sheet 13-2

Metal to be threaded

Tap and tap wrench

Face shield

Protective clothing

Leather gloves

Cutting oil

Vise

Procedure

1. Obtain permission.
2. Put on the face shield.
3. Put on the protective clothing.
4. Put on the leather gloves.
5. Mark the location of the hole to be threaded.
6. Select the proper size drill for the tap size desired (see text Figure 13-9).
7. Drill the pilot hole.
8. Put the tap in the tap wrench.
9. Hold the tap perpendicular to the hole and turn a quarter turn.
10. Reverse to break the scrap.
11. Turn 1 to 2 turns and reverse.
12. Repeat until the threads are completed.
13. Use a taper tap if the hole goes through the metal.
14. Use a plug tap and then a bottoming tap for a hole that does not go through the metal.
15. Remove the tap by turning backward.
16. Remove the chips by wiping with a rag or paper towels.
17. Dispose of the oily rag in the proper container.
18. Clean the area and put away the materials.

Instructor's Score or Approval _____

Job Sheet 13-3

Name _____ Date _____

Threading a Rod or Bolt

Objective

When you have completed this activity you will have demonstrated the ability to thread a rod or bolt.

Tools and Materials

> Job sheet 13-3
> Rod or bolt to be threaded
> Die and Die holder
> Vise
> Face shield
> Cutting oil
> Protective clothing
> Leather gloves
> File
> Wiping rags or paper towels

Procedure

1. Obtain permission.
2. Put on the face shield.
3. Put on the protective clothing.
4. Put on the leather gloves.
5. Clamp the bolt or rod in the vise.
6. Slightly taper the tip of the rod with the file.
7. Select the die of the desired thread type.
8. Mount the die in the die stock.
9. Put the tapered side over the rod.
10. Use cutting oil and turn 1/2 turn.
11. Back up 1/4 turn.
12. Repeat steps 11 and 12 until a short length is threaded.
13. Remove the die and check the fit of the nut.
14. Continue until the desired length is threaded.
15. Remove the die and clean the chips.
16. Dispose of the oily rag properly.
17. Clean the area and put away the materials.

Instructor's Score or Approval _____

Job Sheet 13-4

Name _____ Date _____

Cutting and Threading Pipe

Objective

When you have completed this activity you will have demonstrated the ability to cut and thread pipe.

Tools and Materials

Job sheet 13-4
Pipe to be cut and threaded
Pipe cutter
Pipe die and stock
Pipe reamer
Face shield
Protective clothing
Leather gloves
Cutting oil
Wiping rags
Pipe vise

Procedure

1. Obtain permission.
2. Put on the face shield.
3. Put on the protective clothing.
4. Put on the leather gloves.
5. Clamp the pipe in the pipe vise.
6. Loosen the Pipe cutter until the cutting wheel slides over the pipe.
7. Slide the cutter to the mark where the pipe is to be cut.
8. Turn the handle of the pipe cutter until the cutting wheel bites into the pipe.
9. Rotate the cutter 1 or 2 times using cutting oil.
10. Tighten the pipe cutter 1/2 turn and repeat step 9.
11. Continue until the pipe is severed.
12. Install the pipe reamer and remove the ridge.
13. Install the proper pipe die in the die stock and place the tapered edge toward the pipe end.
14. Taper the pipe slightly with the file to ease the start.
15. Apply moderate pressure and turn 1/2 turn.
16. Reverse to break the chip and then turn 1 to 2 times using cutting oil.
17. Repeat steps 15 and 16 until the pipe is threaded.
18. Clean the cutter and die.
19. Clean the area and put away the materials.

Instructor's Score or Approval _____

Job Sheet 13-5

Name _____ Date _____

Fastening Metal with Bolts and Nuts

Objective

When you have completed this activity you will have demonstrated the ability to fasten metal with bolts and nuts.

Tools and Materials

    Job sheet 13-5

    Metal to be fastened

    Bolts and nuts

    Face shield

    Protective clothing

    Vise

    Wrenches to fit the bolt

    Lock washers

    Portable drill

    Appropriate sized drill bit

    Center punch

    Ball peen hammer

    Combination square

Procedure

1. Obtain permission.
2. Put on the face shield.
3. Put on the protective clothing.
4. Mark the metal where the bolts are to go.
5. Center punch the marks and drill the holes.
6. Countersink the holes if the bolt is a flat head stove bolt.
7. Insert the bolts.
8. Add lock washers and nuts.
9. Tighten the nuts until the lock washers are flattened.
10. Clean the area and put away the materials.

Instructor's Score or Approval _____

Job Sheet 13-6

Name _____ Date _____

Fastening Metal with Cap Screws

Objective

When you have completed this activity you will have demonstrated the ability to fasten metal with cap screws.

Tools and Materials

Job sheet 13-6

Cap screws

Lock washers

Taper, plug, and bottoming taps

Tap wrench

Portable electric drill

Twist bits of the proper sizes (see text Figure 13-9)

Metal to be fastened

Face shield

Protective clothing

Wrench to fit the cap screw

Vise

Center punch

Ball peen hammer

Cutting oil

Procedure

1. Obtain permission.
2. Put on the face shield.
3. Put on the protective clothing.
4. Drill the holes in the first piece the same size as the cap screws.
5. Drill the second holes with the proper sized bit for the tap to be used. Drill the hole to the desired depth.
6. Tap the second holes with the taper tap using cutting oil.
7. Repeat using the plug tap.
8. Repeat using the bottoming tap.
9. Place lock washers on the cap screws and insert them through the first metal piece.
10. Thread the cap screws into the second piece and tighten until the lock washers are flattened.
11. Clean the area and put away the materials.

Instructor's Score or Approval _____

Job Sheet 13-7

Name _____ Date _____

Fastening Metal with Rivets

Objective

When you have completed this activity you will have demonstrated the ability to fasten metal with rivets.

Tools and Materials

    Job sheet 13-7

    Metal to be fastened

    Suitable rivets

    Drill and bits

    Face shield

    Protective clothing

    Center punch

    Ball peen hammer

    Vise

    Cold chisel

Procedure

1. Obtain permission.
2. Put on the face shield.
3. Put on the protective clothing.
4. Select a rivet the thickness of the metals plus 1 diameter of the rivet.
5. Cut the rivet with the chisel or a hacksaw.
6. Locate and center punch the spots for the holes.
7. Drill the proper size holes.
8. Insert the rivets and place the heads on the vise and peen the end until the riveting is completed.
9. Clean the area and put away the materials.

Instructor's Score or Approval _____

Job Sheet 13-8

Name _____ Date _____

Fastening Metal with Sheet Metal Screws

Objective

When you have completed this activity you will have demonstrated the ability to fasten metal with sheet metal screws.

Tools and Materials

    Job sheet 13-8

    Metal to be fastened

    Sheet metal screws

    Face shield

    Protective clothing

    Punch or drill

    Screwdriver

Procedure

1. Obtain permission.
2. Put on the face shield.
3. Put on the protective clothing.
4. Mark the spot.
5. Punch or drill the holes the core diameter of the screws.
6. Insert the screw and tighten firmly, avoiding stripping the metal.
7. Clean the area and put away the materials.

Instructor's Score or Approval _____

Job Sheet 13-9

Name _____ Date _____

Tinning a Soldering Copper

Objective

When you have completed this activity you will have demonstrated the ability to tin a soldering copper.

Tools and Materials

    Job sheet 13-9

    Soldering copper

    Sal-ammoniac

    Sandpaper

    Wire solder

    Face shield

    Damp cloth

Procedure

1. Obtain permission.
2. Put on the face shield.
3. Put on the protective clothing.
4. Remove burned materials with the sandpaper.
5. Heat the copper until it smokes in sal-ammoniac.
6. Rub the copper in the sal-ammoniac until it is shiny.
7. Unroll 6" of solder and touch it to the copper; if it is too cool, increase the heat.
8. Repeat the steps until the copper is silver colored.
9. Wipe with the damp cloth; a good silver color indicates proper tinning.
10. Clean the area and put away the materials.

Instructor's Score or Approval _____

Job Sheet 13-10

Name _____ Date _____

Soldering Sheet Metal

Objective

When you have completed this activity you will have demonstrated the ability to solder sheet metal.

Tools and Materials

Job sheet 13-10

Sheet metal to be soldered

Soldering copper

Wire solder

Flux

Damp cloth

Face shield

Protective clothing

Sandpaper

Procedure

1. Obtain permission.
2. Put on the face shield.
3. Put on the protective clothing.
4. Clean the areas to be soldered with the sandpaper.
5. Heat the metal with the flat part of the copper.
6. Add flux as it heats.
7. Touch the solder to the metal and as it flows, move the copper along, following with the solder.
8. Tin both pieces of metal.
9. Place the inner surfaces together and support them as necessary.
10. Heat the pieces to melt the solder.
11. Add solder to fill the metal.
12. Wipe off the excess solder.
13. Hold the metal together until the solder hardens.
14. Clean the area and put away the materials.

Instructor's Score or Approval _____

Job Sheet 13-11

Name _____ Date _____

Soldering Electrical Connections

Objective

When you have completed this activity you will have demonstrated the ability to solder electrical connections.

Tools and Materials

    Job sheet 13-11

    Soldering copper

    Wire connections to be soldered

    Vise

    Pliers

    Electrical tape

    Rosin core solder

    Face shield

    Protective clothing

Procedure

1. Obtain permission.
2. Put on the face shield.
3. Put on the protective clothing.
4. Twist the wires together.
5. Clamp the wires in a vise.
6. Hold the heated copper under the wires and feed the solder from the top.
7. Allow the solder to cool and wrap with electrical tape.
8. Clean the area and put away the materials.

Instructor's Score or Approval _____

Job Sheet 13-12

Name _____ Date _____

Soldering Copper Pipes

Objective

When you have completed this activity you will have demonstrated the ability to solder copper pipes.

Tools and Materials

Job sheet 13-12

Copper pipes

Solder approved for copper pipe

Fine sandpaper

Flux

Gas torch

Face shield

Protective clothing

Procedure

1. Obtain permission.
2. Put on the face shield.
3. Put on the protective clothing.
4. Polish the pipe and the inside of the fitting with the sandpaper.
5. Apply the paste flux.
6. Press the pipe and fitting together.
7. Light the torch.
8. Heat the underside of the fitting with the torch.
9. Hold the solder at the joint between the pipe and the fitting.
10. Heat the pipe until the solder melts and is drawn into the joint and is filled.
11. Remove the heat.
12. Cool the joint.
13. Pressure the system and check for leaks.
14. Clean the area and put away the materials.

Instructor's Score or Approval _____

SECTION 4 POWER TOOLS IN THE AGRICULTURAL MECHANICS SHOP

Unit 14 Portable Power Tools

Portable power tools are relatively inexpensive today and are in common use by home owners as well as by tradesmen and professionals. Learning how to use them safely will be valuable skills for you to know and practice. This unit emphasizes safety in their use.

Shop Activity 14-1

Safety Precautions

Shop Activity 14-2

Using a Power Drill

Shop Activity 14-3

Using a Portable Sander

Shop Activity 14-4

Using a Portable Disc Sander or Grinder

Shop Activity 14-5

Using a Finishing Sander

Shop Activity 14-6

Using a Portable Saber Saw

Shop Activity 14-7

Using a Reciprocating Saw

Shop Activity 14-8

Using a Circular Saw

Shop Activity 14-9

Using a Power Router

Job Sheet 14-1

Name _____ Date _____

Safety Precautions

Objective

When you have completed this activity you will have demonstrated the ability to identify portable power tool safety precautions.

Tools and Materials

Job sheet 14-1

Pen or Pencil

Clipboard

Procedure

Complete the checklist for the portable power tools in the shop. If you find a tool that does not meet the checklist, notify the instructor and do not use it until the condition is corrected.

1. All power cords on metal housed tools are 3-wire safety.
2. All power cords are free from breaks and cuts.
3. All power cords are properly attached to tools.
4. Blades on cutting tools are clean and sharp.
5. Tools are clean and dry.
6. Tools are not used in wet areas.
7. Protective clothing is worn when tools are used.
8. Work is supported when using portable power tools.
9. Power tools are stored in a manner that protects the cutting edges.
10. Permission of the instructor is required before a tool is used.
11. Power tools are not forced or overloaded.
12. Others are warned when tools are to be used.
13. Vent holes in the tools are free from dirt.
14. All switches work properly.
15. All guards operate properly.

Instructor's Score or Approval _____

Job Sheet 14-2

Name _____ Date _____

Using a Power Drill

Objective

When you have completed this activity you will have demonstrated the ability to use a portable power drill.

Tools and Materials

- Job sheet 14-2
- Portable power drill
- Twist bit
- File
- Center punch
- Combination square
- Soapstone marker
- Ball peen hammer
- Metal to be drilled
- Face shield
- Leather gloves
- Protective clothing
- Cutting oil

Procedure

1. Obtain permission.
2. Put on the face shield.
3. Put on the protective clothing.
4. Put on the leather gloves.
5. Mark the drilling location in the metal.
6. Center punch the spot with a gentle tap.
7. Check the location and correct if necessary, and give the center punch one good blow.
8. Tighten the drill bit in the chuck using at least two holes.
9. Remove the chuck wrench.
10. Hold the drill perpendicular to the metal and start the cut.
11. Use moderate pressure.
12. Use cutting oil to cool the bit.
13. Lower the pressure as the bit breaks through.
14. Remove any burrs with a file.
15. Clean the area and put away the materials.

Instructor's Score or Approval _____

Job Sheet 14-3

Name _____ Date _____

Using a Portable Sander

Objective

When you have completed this activity you will have demonstrated the ability to use a portable sander.

Tools and Materials

    Portable sander

    Dust mask

    Face shield

    Protective clothing

    Wood to be sanded

    Appropriate sandpaper belts

    Holding device, vise, or bench stops

Procedure

1. Obtain permission.
2. Put on the dust mask.
3. Put on the face shield.
4. Put on the protective clothing.
5. Clamp the wood to be sanded.
6. Select the appropriate belt and install it.
7. Empty the dust bag if needed.
8. Hold the machine above the work and turn it on.
9. Hold the power cord out of the path of the sander.
10. Lower the sander and keep the sander moving over the work, sanding with the grain.
11. Raise the sander and turn it off.
12. Check the work and repeat as necessary to obtain an even smooth surface.
13. Install a fine belt and sand to finish.
14. Empty the dust bag.
15. Clean the area and put away the materials.

Instructor's Score or Approval _____

Job Sheet 14-4

Name _____ Date _____

Using a Portable Disc Sander or Grinder

Objective

When you have completed this activity you will have demonstrated the ability to use a portable disc sander or grinder.

Tools and Materials

    Object to be sanded

    Portable disc sander or grinder

    Face shield

    Dust mask

    Protective clothing

    Vise or clamps

Procedure

1. Obtain permission.
2. Put on the dust mask.
3. Put on the face shield.
4. Put on the protective clothing.
5. Clamp the metal to be sanded.
6. Select the appropriate disc or grinding wheel.
7. Hold the machine in both hands and turn on the switch.
8. Lower to the work slowly and let the weight of the machine be the only pressure.
9. Keep the machine moving.
10. Lift before turning off the machine and hold it until it stops turning.
11. Check the work and repeat sanding or grinding as needed.
12. Lay the machine on the proper flat spot and not on the disc or grinding wheel.
13. Clean the area and put away the materials.

Instructor's Score or Approval _____

Job Sheet 14-5

Name _____ Date _____

Using a Finishing Sander

Objective

When you have completed this activity you will have demonstrated the ability to use a finishing sander.

Tools and Materials

Job sheet 14-5

Object to be sanded

Dust mask

Face shield

Protective clothing

Finishing sander

Sandpaper

Procedure

1. Obtain permission.
2. Put on the dust mask.
3. Put on the face shield.
4. Put on the protective clothing.
5. Clamp the object to be sanded.
6. Select the appropriate sandpaper and install it.
7. Start the sander and keep it moving, using coarser papers first.
8. As sanding proceeds, switch to finer sandpaper.
9. Remove dust frequently to prevent clogging the sandpaper.
10. Clean the area and put away the materials.

Instructor's Score or Approval _____

Job Sheet 14-6

Name _____ Date _____

Using a Portable Saber Saw

Objective

When you have completed this activity you will have demonstrated the ability to use a portable saber saw.

Tools and Materials

    Job sheet 14-6

    Portable saber saw

    Material to be cut

    Dust mask

    Face shield

    Protective clothing

    Vise or clamps

Procedure

1. Obtain permission.
2. Put on the dust mask.
3. Put on the face shield.
4. Put on the protective clothing.
5. Clamp the work.
6. Select and install the appropriate blade.
7. Set the angle required.
8. Start the cut at an edge or bore a hole for an inside cut.
9. In a blind cut put the toe of the base on the work with the blade free and rotate the saw until the cut is started.
10. Use firm pressure and steady movement to cut, allowing time for the blade to cut in turns and curves.
11. When finishing a cut, lessen the pressure and cutting speed.
12. Remove the blade and lightly tighten the blade-holding screw.
13. Clean the area and put away the materials.

Instructor's Score or Approval _____

Job Sheet 14-7

Name _____ Date _____

Using a Reciprocating Saw

Objective

When you have completed this activity you will have demonstrated the ability to use a reciprocating saw.

Tools and Materials

    Job sheet 14-7

    Reciprocating saw

    Vise or clamps

    Dust mask

    Face shield

    Combination square

    Pencil

    Protective clothing

    Material to be cut

Procedure

1. Obtain permission.
2. Put on the dust mask.
3. Put on the face shield.
4. Put on the protective clothing.
5. Clamp the material to be cut.
6. Mark the cut.
7. Select and install the correct blade for the job.
8. Select the proper cutting speed.
9. Keep the shoe against the work at all times.
10. Use the same techniques as the saber saw while cutting.
11. Remove the blade when finished.
12. Clean the area and put away the materials.

Instructor's Score or Approval _____

Job Sheet 14-8

Name _____ Date _____

Using a Circular Saw

Objective

When you have completed this activity you will have demonstrated the ability to use a circular saw.

Tools and Materials

Job sheet 14-8

Portable circular saw

Dust mask

Face shield

Protective clothing

Wood to be cut

Vise or clamps

Procedure

1. Obtain permission.
2. Put on the dust mask.
3. Put on the face shield.
4. Put on the protective clothing.
5. Clamp the material.
6. Check the blade for the job.
7. Adjust the desired angle and the depth of cut.
8. If ripping, set the rip guide for the desired width so the blade cuts in the waste.
9. If crosscutting, mark the cut and saw on the waste side of the mark.
10. If a pocket cut is to be made, use the front of the base with the blade free and lower the blade slowly into the work.
11. Hold the saw until the blade stops turning before setting it down.
12. Clean the area and put away the materials.

Instructor's Score or Approval _____

Job Sheet 14-9

Name _____ Date _____

Using a Power Router

Objective

When you have completed this activity you will have demonstrated the ability to use a power router.

Tools and Materials

Job sheet 14-9

Portable router

Appropriate router bit

Appropriate jig or guide

Extension cord

Dust mask

Face shield

Protective clothing

Scrap of test material

Material to be shaped

Vise or clamps

Procedure

1. Obtain permission.
2. Put on protective clothing.
3. Put on the dust mask.
4. Select and install the proper router bit.
5. Adjust the depth of cut.
6. Clamp the test scrap securely.
7. Plug in the router.
8. Place the base plate of the router on the test material near the starting point of the cut.
9. Turn on the router and make the test cut moving the router slowly enough to allow the router to maintain speed without glazing the material.
10. Turn off the router and place it so it cannot roll.
11. Check the cut for proper shape and depth of cut.
12. Readjust and repeat the test as necessary.
13. Clamp the material to be shaped securely.
14. Turn on the router and make the cut following the guide closely.
15. Finish the cut and turn off the router.
16. Remove the bit.
17. Clean and store the router and the bit.
18. Clean the area and have your instructor inspect your work.

Instructor's Score or Approval _____

UNIT 15 Woodworking with Power Machines

Stationary power machines are very important in modern construction. They have certain characteristics that make them more valuable than the portable power tools in the previous unit. Learning to use them properly and safely will be valuable to you.

Class Activity 15-1

Identifying Safety Precautions with Power Woodworking Machines

Shop Activity 15-2

Cutting with the Band Saw

Shop Activity 15-3

Cutting with the Jigsaw

Shop Activity 15-4

Cutting with the Table Saw

Shop Activity 15-5

Checking the Radial Arm Saw for Levelness

Shop Activity 15-6

Crosscutting with the Radial Arm Saw

Shop Activity 15-7

Ripping with the Radial Arm Saw

Shop Activity 15-8

Using a Cutoff Saw

Shop Activity 15-9

Using the Jointer to Smooth an Edge

Shop Activity 15-10

Planing a Board to Thickness

Shop Activity 15-11

Smoothing a Board with a Stationary Sander

Job Sheet 15-1

Name _____ Date _____

Identifying Safety Precautions with Power Woodworking Machines

Objective

When you have completed this activity you will have demonstrated the ability to identify safety precautions with power woodworking machines.

Tools and Materials

Job sheet 15-1

Clipboard

Pen or pencil

Procedure

Complete the safety checklist by matching the items in the two columns.

_____ 1. Machine use   a. before leaving
_____ 2. Safety zone   b. before changing blades
_____ 3. Push sticks   c. for large pieces
_____ 4. Unplug machine   d. brush or vacuum
_____ 5. Cleaning machines   e. for short pieces
_____ 6. Slow and cautious   f. wear in the shop
_____ 7. Turn off machine   g. only the operator allowed
_____ 8. Overworked machine   h. good working condition
_____ 9. Get help   i. obtain permission
_____ 10. Goggles or face shield j. all operations
_____ 11. Protective clothing   k. you have had training for
_____ 12. Only procedures   l. around machines
_____ 13. Walk   m. not loose or baggy
_____ 14. Only use machines in   n. overload protection

Instructor's Score or Approval _____

Job Sheet 15-2

Name _____ Date _____

Cutting with the Band Saw

Objective

When you have completed this activity you will have demonstrated the ability to cut with the band saw.

Tools and Materials

    Job sheet 15-2

    Band saw

    Face shield or goggles

    Shop coat or apron

    Wood to be cut

Procedure

1. Obtain the instructor's permission.
2. Put on face shield and protective clothing.
3. Check that the blade is proper for the job.
4. Check the blade adjustment.
5. Set the guide assembly to within 1/8″ of the stock.
6. Set the miter gauge to the desired angle if it is used.
7. Set the rip fence to the desired measurement if it is to be used.
8. Clear the table of any material.
9. Stand aside and start the machine.
10. Check that the blade is tracking properly.
11. Slowly push the board into the blade.
12. For straight cuts move the board in a straight line.
13. For curved cuts rotate the board slowly as the saw cuts.
14. Plan cuts so they go all the way through the board.
15. Complete the cuts.
16. Shut off the saw.
17. Make sure the blade stops before leaving the machine.
18. Clean the area and put away the materials.

Instructor's Score or Approval _____

Job Sheet 15-3

Name _____ Date _____

Cutting with the Jigsaw

Objective

When you have completed this activity you will have demonstrated the ability to cut wood with the jigsaw.

Tools and Materials

    Job sheet 15-3

    Jigsaw

    Wood

    Face shield or goggles

    Shop coat or apron

Procedure

1. Obtain permission to use the saw.
2. Put on the face shield or goggles.
3. Put on the protective clothing.
4. Select a suitable blade.
5. Unplug the motor.
6. Attach the blade to the lower chuck.
7. Rotate the motor pulley to the blade's highest point.
8. Attach the blade to the upper chuck.
9. Loosen the tension sleeve clamp knob.
10. Lift the tension sleeve until there is moderate tension on the blade.
11. Tighten the tension sleeve clamp knob.
12. Check that the blade goes all the way up without buckling and readjust if necessary.
13. Adjust the hold down foot tension to slight tension on the work.
14. Plug the machine back in and turn it on.
15. Keep fingers to the side and push the wood slowly.
16. Saw on the waste side of the line.
17. Rotate the stock so the saw blade follows the line.
18. Cut out through waste and don't back out of cuts.
19. To cut inside a circle, bore a hole through the waste material and insert the blade.
20. Saw inside the circle.
21. Remove the blade.
22. Clean the area and put away the materials.

Instructor's Score or Approval _____

Job Sheet 15-4

Name _____ Date _____

Cutting with the Table Saw

Objective

When you have completed this activity you will have demonstrated the ability to cut with a table saw.

Tools and Materials

Job sheet 15-4

Table Saw

Face shield or goggles

Shop coat or apron

Wood to be cut

T-bevel

Procedure

1. Obtain permission to use the saw.
2. Put on the face shield or goggles.
3. Put on the protective clothing.
4. Unplug the motor.
5. Install the correct blade for the job.
6. Adjust the blade angle with the scale or a T-bevel.
7. Adjust the blade height so only the teeth extend beyond the stock.
8. If ripping, remove the miter gauge from the saw.
9. Measure from the tip of the tooth nearest the rip fence to set the desired width.
10. If crosscutting, remove the rip fence or move it out of the way.
11. Set the miter gauge to the desired angle.
12. Check that the guard assembly is in place.
13. Plug in the motor.
14. Stand aside and start the machine.
15. Push the material slowly toward the blade.
16. Hold the material firmly on both sides of the blade.
17. Use a push stick to complete the cut.
18. Shut off the machine.
19. Clean the area and put away the materials.

Instructor's Score or Approval _____

Job Sheet 15-5

Name _____ Date _____

Checking the Radial Arm Saw Table for Levelness

Objective

After completing this activity you will have demonstrated the ability to check the radial arm saw table for levelness.

Tools and Materials

Job sheet 15-5

Radial arm saw

Procedure

1. Obtain the instructor's permission.
2. Unplug the motor.
3. Loosen the column clamp.
4. Turn the elevating crank until the teeth just touch the table.
5. Loosen the trolley lock.
6. Pull the saw across the table and if the teeth don't lightly touch the table, report the condition to the instructor and do not use the saw until the condition is corrected.

Instructor's Score or Approval _____

Job Sheet 15-6

Name _____ Date _____

Crosscutting with the Radial Arm Saw

Objective

When you have completed this activity you will have demonstrated the ability to crosscut with the radial arm saw.

Tools and Materials

Job sheet 15-6

Radial arm saw

Face shield or goggles

Shop coat or apron

Wood to be cut

Procedure

1. Obtain the instructor's permission.
2. Put on the face shield.
3. Put on the protective clothing.
4. Mark the point of the cut.
5. Set the saw to just cut through the board.
6. Set the board so the saw will cut on the waste side of the mark.
7. Hold the saw with one hand and turn it on.
8. Hold with the other hand not in line with the blade.
9. Slowly pull the saw through the board.
10. Return the saw to the column.
11. Turn off the saw and hold it until it stops turning.
12. Clean the area and put away the materials.

Instructor's Score or Approval _____

Job Sheet 15-7

Name _____ Date _____

Ripping with the Radial Arm Saw

Objective

Upon completing this activity you will have demonstrated the ability to rip with the radial arm saw.

Tools and Materials

Job sheet 15-7

Radial arm saw

Wood to be ripped

Face shield or goggles

Shop coat or apron

Pencil

Combination square

Procedure

1. Obtain permission.
2. Put on the face shield.
3. Put on the shop coat or apron.
4. Elevate the saw until the teeth clear the table.
5. Release the yoke lock and turn the saw 90 degrees.
6. Measure the desired width of the cut by measuring to an inside tooth.
7. Lock the trolley in this position.
8. Lower the blade until the saw just touches the table.
9. Rotate the nose of the blade guard so it just clears the work by 1/8".
10. Lower the antikickback device to the height of the work.
11. Obtain a helper for a long board.
12. Stand aside and turn on the saw.
13. Push the wood slowly through the saw.
14. For short boards finish with a push stick.
15. Turn off the saw and wait until it stops turning.
16. Elevate the saw and return it to the crosscutting position.
17. Clean the area and put away the materials.

Instructor's Score or Approval _____

Job Sheet 15-8

Name _____ Date _____

Using a Cutoff Saw

Objective

When you have completed this activity you will have demonstrated the ability to use a cutoff saw.

Tools and Materials

    Job sheet 15-8

    Cutoff saw

    Appropiate blade for the saw

    Face shield

    Dust mask

    Protective clothing

    Material to be cut

    Extension cord

Procedure

1. Obtain permission to use the cutoff saw.
2. Put on protective clothing.
3. Put on dust mask.
4. Put on the face shield.
5. Unplug the machine and install the appropriate blade.
6. Set the saw for the desired angle of cut.
7. Plug the machine back into the outlet or extension cord.
8. Stand to the side and start the machine, holding with one hand while turning the switch.
9. Make the cut lowering the saw slowly into and through the stock.
10. Allow the saw to slowly rise to the parked position.
11. Turn off the saw and hold it until it stops turning.
12. Remove the stock.
13. Clean the area.

Instructor's Score or Approval _____

Job Sheet 15-9

Name _____ Date _____

Using the Jointer to Smooth an Edge

Objective

When you have completed this activity you will have demonstrated the ability to smooth an edge of a board with the jointer.

Tools and Materials

    Job sheet 15-9

    Jointer

    Board to smooth

    Face shield or goggles

    Protective clothing

Procedure

1. Obtain permission to use the jointer.
2. Put on the face shield.
3. Put on the protective clothing.
4. Check the blade guard for proper operation.
5. Adjust the fence to the desired angle.
6. Adjust the depth of cut to 1/16".
7. Plane only boards over 12" in length.
8. Obtain a helper if the board is over 48" long.
9. Stand aside and start the machine.
10. Start the board on the front feed table holding it by the top edge.
11. Move forward slowly with moderate down pressure.
12. Move the hands back as they near the blades.
13. Move the second hand over the rear table as the end nears the blade.
14. Continue lowering the pressure as the cut ends.
15. Use a push stick for finishing a board under 48".
16. Reverse the board to keep it straight.
17. Make the last pass with the grain.
18. Clean the area and put away the materials.

Instructor's Score or Approval _____

Job Sheet 15-10

Name _____ Date _____

Planing a Board to Thickness

Objective

When you have completed this activity you will have demonstrated the ability to plane a board to thickness.

Tools and Materials

    Job sheet 15-10

    Face shield

    Protective clothing

    Wood to be planed

Procedure

1. Obtain permission.
2. Put on the face shield.
3. Put on the protective clothing.
4. Check the machine guard operation.
5. Identify the thickest board end.
6. Measure the thickness and set the depth of cut 1/8" less (depth may vary with the hardness of the wood).
7. Select a helper when required.
8. Place the thinner end on the infeed table.
9. Stand aside and turn on the machine.
10. Push the board until the feed rollers take over.
11. On long boards, move the helper to the rear at the halfway point.
12. Plane any additional boards.
13. Raise the bed and make the second pass on the opposite side.
14. Make shallower passes to finish with the grain.
15. Turn off the machine and stand by until it stops.
16. Clean the area and put away the materials.

Instructor's Score or Approval _____

Job Sheet 15-11

Name _____ Date _____

Smoothing a Board with a Stationary Sander

Objective

When you have completed this activity you will have demonstrated the ability to smooth a board with a stationary sander.

Tools and Materials

    Job sheet 15-11

    Board to be sanded

    Face shield

    Protective clothing

    Dust mask

    Stationary sander

Procedure

1. Obtain permission.
2. Put on the face shield.
3. Put on the protective clothing.
4. Put on the dust mask.
5. Put an unclogged belt on the sander.
6. Turn on the dust collector.
7. Turn on the sander.
8. Hold the work firmly with fingers away from the belt.
9. Put the wood lightly on the belt.
10. Move the wood back and forth sideways.
11. On the disc, use the portion moving down.
12. Use only moderate pressure.
13. Reduce the pressure before removing the wood.
14. In freehand sanding, use light pressure and keep the work moving at all times.
15. Turn off the sander and dust collector.
16. Wait until the sander stops.
17. Clean the area and put away the materials.

Instructor's Score or Approval _____

Unit 16 Metalworking with Power Machines

The power machines covered in this unit are the drill press, grinder, power hacksaw, metal cutting shear, and the metal break.

Shop Activity 16-1

Boring Holes with the Drill Press

Shop Activity 16-2

Shaping Metal with the Grinding Wheel

Shop Activity 16-3

Cleaning Metal with the Wire Wheel

Shop Activity 16-4

Dressing the Grinding Wheel

Shop Activity 16-5

Cutting Stock to Length with the Power Hacksaw

Shop Activity 16-6

Cutting Mild Steel with the Power Shear

Job Sheet 16-1

Name _____ Date _____

Boring Holes with the Drill Press

Objective

When you have completed this activity you will have demonstrated the ability to bore holes with the drill press.

Tools and Materials

    Job sheet 16-1
    Face shield
    Protective clothing
    Leather gloves
    Drill bit
    Metal to be drilled
    Center punch
    Ball peen hammer
    Combination square
    Cutting oil
    Chuck wrench
    Metal file
    Drill press vise

Procedure

1. Obtain permission to use the machine.
2. Put on the face shield.
3. Put on the protective clothing.
4. Put on the leather gloves.
5. With the combination square, mark the spot to be drilled.
6. Place the center punch on the mark to be drilled and give it a light tap.
7. Check the mark for accuracy.
8. Correct if needed and give the center punch one sharp blow.
9. Place the metal in the drill press vise or clamp securely.
10. Use a pilot drill if the hole is to be over 3/8" in diameter.
11. Install the drill and tighten by using two holes.
12. Remove the chuck wrench.
13. Turn on the machine.
14. Lower the feed handle and start the hole using steady pressure.
15. Add cutting oil to cool the bit.
16. Reduce pressure as the bit breaks through.
17. Remove any burrs with a file.
18. Clean the area and put away the materials.

Instructor's Score or Approval _____

Job Sheet 16-2

Name _____ Date _____

Shaping Metal with the Grinding Wheel

Objective

When you have completed this activity you will have demonstrated the ability to shape metal with a grinding wheel.

Tools and Materials

    Job sheet 16-2

    Face shield

    Protective clothing

    Leather gloves

    Metal to be shaped

    Container of water

    Grinding wheel

Procedure

1. Obtain permission.
2. Put on the face shield.
3. Put on the protective clothing.
4. Put on the leather gloves.
5. Check the wheel using only those with no nicks or chips.
6. Adjust the tool rest to about 1/16" clearance.
7. Set the tool rest to the center of the shaft.
8. If bevel grinding, set the appropriate angle.
9. Hold the metal firmly on the tool rest.
10. Stand to the side and turn on the grinder.
11. Move back and forth in the needed shaping pattern.
12. Cool the metal as needed with the water.
13. Clean the area and put away the material.

Instructor's Score or Approval _____

Job Sheet 16-3

Name _____ Date _____

Cleaning Metal with the Wire Wheel

Objective

When you have completed this activity you will have demonstrated the ability to clean metal with a wire wheel.

Tools and Materials

  Job sheet 16-3

  Face shield

  Protective clothing

  Leather gloves

  Metal to be cleaned

  Wire wheel

  Dust mask

Procedure

1. Obtain permission.
2. Put on the face shield.
3. Put on the protective clothing.
4. Put on the leather gloves.
5. Put on the dust mask.
6. Stand aside and turn on the machine.
7. Hold the metal firmly in the hands.
8. Move the metal back and forth using moderate pressure until it is clean.
9. Clean the area and put away the materials.

Instructor's Score or Approval _____

Job Sheet 16-4

Name _____ Date _____

Dressing the Grinding Wheel

Objective

When you have completed this activity you will have demonstrated the ability to dress the grinding wheel.

Tools and Materials

    Job sheet 16-4

    Face shield

    Protective clothing

    Leather gloves

    Grinding wheel

    Dust mask

    Wheel dresser

    Combination square

Procedure

1. Obtain permission.
2. Put on the face shield.
3. Put on the protective clothing.
4. Put on the leather gloves.
5. Put on the dust mask.
6. Stand aside and turn on the machine.
7. Hold the wheel dresser square with the face of the grinding wheel.
8. Move the dresser back and forth slowly for 30 seconds to clean and true the wheel.
9. Readjust the tool rest to 1/16".
10. Rotate the wheel by hand to check the roundness.
11. Use the combination square to check squareness of the face.
12. Repeat dressing if needed.
13. Clean the area and put away the materials.

Instructor's Score or Approval _____

Job Sheet 16-5

Name _____ Date _____

Cutting Stock to Length with the Power Hacksaw

Objective

When you have completed this activity you will have demonstrated the ability to cut stock to length with the power hacksaw.

Tools and Materials

    Job sheet 16-5

    Face shield

    Protective clothing

    Leather gloves

    Stock to be cut

    Power hacksaw

    Combination square

Procedure

1. Obtain permission.
2. Put on the face shield.
3. Put on the protective clothing.
4. Put on the leather gloves.
5. With the square, mark the metal where it is to be cut.
6. Check the blade on the machine to match the cut to be made.
7. Raise the saw blade.
8. Set the material in the vise at the desired spot so the blade cuts in the waste.
9. Turn on the machine and lower the blade slowly.
10. Turn on the coolant.
11. Stand by while the cut is made.
12. Turn off the machine when finished.
13. Remove the metal and the scrap.
14. Clean the area and put away the materials.

Instructor's Score or Approval _____

Job Sheet 16-6

Name _____ Date _____

Cutting Mild Steel with the Power Shear

Objective

When you have completed this activity you will have demonstrated the ability to cut mild steel with the power shear.

Tools and Materials

    Job sheet 16-6

    Face shield

    Protective clothing

    Leather gloves

    Power shear

    Combination square

    Metal to be cut

Procedure

1. Obtain permission.
2. Put on the face shield.
3. Put on the protective clothing.
4. Put on the leather gloves.
5. With the combination square, mark the metal.
6. Raise the shear handle.
7. Align the metal so the shear will cut in the waste.
8. Support the metal level.
9. Lower the handle slowly and safely.
10. Remove the metal and the scrap.
11. Clean the area and put away the materials.

Instructor's Score or Approval _____

SECTION 5 PROJECT PLANNING

Unit 17 Sketching and Drawing Projects

The time one spends in planning is amply rewarded in the construction phase. Learning how to sketch and draw will help you plan the details of a project.

A sketch or a drawing is a representation of the project that shows the visible features of the object so one can tell how it should look. These also help you determine the amounts and kinds of materials that are needed to make the object.

Class Activity 17-1

Identifying the Commonly Used Drawing Materials

Class Activity 17-2

Identifying the Commonly Used Lines in Sketching or Drawing

Class Activity 17-3

Understanding Sketches and Drawings

Class Activity 17-4

Practicing the Use of a Scale

Class Activity 17-5

Making a Three-View Drawing

Job Sheet 17-1

Name _____ Date _____

Identifying the Commonly Used Drawing Materials

Objective

When you have completed this activity you will have demonstrated the ability to identify the commonly used drawing materials.

Tools and Materials

Job sheet 17-1

Pen or pencil

Procedure

Complete the checklist by matching the object with its use.

    ___ 1. Measure or draw angles   a. T-square

    ___ 2. Draw curved lines   b. triangle

    ___ 3. Draw parallel lines   c. protractor

    ___ 4. Makes the lines   d. eraser

    ___ 5. Corrects mistakes   e. pencil

    ___ 6. Proportional drawing   f. drawing board

    ___ 7. Drawing surface   g. scale

    ___ 8. 90 degree or angled lines   h. ruler

    ___ 9. Measure lines   i. compass

Instructor's Score or Approval _____

Job Sheet 17-4

Name _____ Date _____

Practicing the Use of a Scale

Objective

Upon completion of this activity you will have demonstrated the ability to use a scale.

Tools and Materials

Job sheet 17-4

Pencil

Scale

Procedure

In the space provided, draw the desired lines to the scale indicated.

Scale	Line
1/8" = 1'	3"
1/4" = 1'	5'
3/16" = 1'	8'

Instructor's Score or Approval _____

Job Sheet 17-5

Name _____ Date _____

Making a Three-View Drawing

Objective

When you have completed this activity you will have demonstrated the ability to make a three-view drawing.

Tools and Materials

    Job sheet 17-5

    4-H pencil

    Eraser

    Scale

    T-square

    Triangle

    Drafting paper

    Masking tape

    Drawing board

Procedure

Because this is your first drawing experience, we will have you repeat the three-view drawing in the text (Figure 17-17 from text) with a different scale, allowing you to follow the same procedure.

1. Place the drafting paper on the drawing board using the T-square to align it with the board.
2. Tape the corners to the drawing board.
3. Draw a heavy line 1" from each edge of the paper for the border lines.
4. Measure 1/2" up from the bottom border line and draw a heavy line for the title block.
5. With the T-square and triangle, divide the title block at 2 1/2" from each end and at 4 1/2" from the left end.
6. Letter your name, date, show box, and scale 1 1/2" = 1' in the title blocks.
7. Make a light mark 2" from the left border line and draw a very light vertical line from border to border line.
8. Measure to the right 6" and repeat.
9. Measure 2" and 5" to the right and draw very light lines 4" up from the title block.
10. Measure 2" up from the title block, and draw a light line from the first vertical line to the second, and from the third to the fourth parallel to the title block.
11. Measure 4 1/4" up from the title block and repeat the lines as in step 10.
12. Measure 6 1/4" up from the title block and draw a light line between the first two vertical lines.
13. Measure 9 1/4" up from the title block and draw a light line between the first two vertical lines completing the top and bottom object lines.
14. Measure 1/8" in from the left and right sides of the bottom view and draw a broken vertical line through the bottom and top views.
15. Measure 1/8" up from the bottom view and draw a broken horizontal line through the bottom view and the broken lines drawn previously.
16. Continue this broken line through the end view.

17. Measure down from the top of the front view and draw a solid horizontal line from solid line to solid line to represent the lid, erasing any broken lines extending into this space.
18. Repeat step 17 for the end view.
19. Measure 1/8" up from the bottom of the top view and draw a broken line between the two broken lines drawn in step 14.
20. Measure down from the top of the top view and repeat step 19.
21. Measure 1/16" from the bottom of the top view and draw a broken line from the left and right object lines to represent the 3/8" rabbet.
22. Erase any broken lines drawn in step 14 that extend beyond the line drawn in step 21.
23. Measure 1/16" down from the top of the top view and repeat steps 21 and 22.
24. Measure 1/8" in from the left and right sides of the end view and draw a vertical broken line between the bottom and the lid.
25. Erase any broken lines extending beyond these two broken lines.
26. Measure 1/16" in from the left and right sides of the end view and draw a solid line from the bottom to the lid to represent the visible end of the rabbet joint.
27. Measure 1 1/2" up from the bottom and mark where the handles are to be attached.
28. Draw a horizontal line 1/8" extending from the left and right sides of the front view.
29. Locate the center of the top view and measure 3/8" up and down from the center and draw 1/8" horizontal lines from each side of the top.
30. With the T-square and triangle, complete the vertical lines between the lines in step 28.
31. Draw 1 1/4" vertical lines up from the lines drawn in step 27.
32. Measure 1/4" up from the mark established in step 26.
33. Draw a line from this mark to the line completed in step 30.
34. Locate the center of the end view and measure 3/8" each way from the center and extend the lines from the front view between the two marks.
35. Complete the end view of the handles by drawing the vertical lines.
36. Clean the area and put away the materials.

Instructor's Score or Approval _____

Unit 18 Figuring a Bill of Materials

Before you can build a project you must obtain the materials needed for construction. Knowing how to figure a bill of materials is an essential skill in project construction. In this unit you will learn how to use the drawings created in the last unit to figure a bill of materials.

Class Activity 18-1

Identifying the Components of a Bill of Materials

Class Activity 18-2

Figuring the Bill of Materials for the Show Box

Class Activity 18-3

Recognizing Common Units of Measure

Class Activity 18-4

Calculating Board Feet

Class Activity 18-5

Identifying How Materials Are Priced

Job Sheet 18-1

Name _____ Date _____

Identifying the Components of a Bill of Materials

Objective

Upon completing this activity you will have demonstrated the ability to identify the components of a bill of materials.

Tools and Materials

Job sheet 18-1

Pen or pencil

Procedure

Complete the following worksheet.

1. What is the bill of materials?
   _____
   _____

2. What are the items of information contained in the bill?
   _____
   _____

3. What is a board foot of lumber?
   _____

4. What are the two methods of computing board feet?
   _____
   _____
   _____

Instructor's Score or Approval _____

Job Sheet 18-2

Name _____ Date _____

Figuring the Bill of Materials for the Show Box

Objective

When you have completed this activity you will have demonstrated the ability to figure a simple bill of materials.

Tools and Materials

    Job sheet 18-2

    Pen or pencil

    Drawing in text (Figure 17-17)

Procedure

Complete the following materials list form.

Item	Number	Size	Bd. Ft.	Cost*
Top	____	_____	_____	$ _____
Bottom	____	_____	_____	_____
Sides	____	_____	_____	_____
Ends	____	_____	_____	_____
Handles	____	_____	_____	_____
Hinges	____	_____	(Not shown)	_____
Hasp	____	(Not shown, optional)		_____
Finish	____			_____
*Totals***			_____	$ _____

    * Obtain the prices from your instructor

    ** Nails, screws, and glue are not included

Instructor's Score or Approval _____

Job Sheet 18-3

Name _____ Date _____

Recognizing Common Units of Measure

Objective

When you have completed this activity you will have demonstrated the ability to recognize the common units of measure.

Tools and Materials

Job sheet 18-3

Pen or pencil

Procedure

Complete the following worksheet.

____ 1. Lumber   a. pound

____ 2. Pipe   b. thousand board feet

____ 3. Plywood   c. sheet or board foot

____ 4. Steel sheets or rounds and angles   d. standard length of 21"

Instructor's Score or Approval _____

Job Sheet 18-4

Name _____ Date _____

Calculating Board Feet

Objective

When you have completed this activity you will have demonstrated the ability to calculate board feet.

Tools and Materials

    Job sheet 18-4

    Pen or pencil

Procedure

Calculate the number of board feet in the following list of lumber materials for a hayrack.

Item	Number	Size	Bd. Ft. Each	Total Bd. Ft.
Sills	2	2″ x 8″ x 14′	_____	_____
Joists	8	2″ x 6″ x 8′	_____	_____
Side Rails	2	2″ x 4″ x 14′	_____	_____
Floor	16	1″ x 6″ x 14′	_____	_____
Standards	4	2″ x 4″ x 6′	_____	_____
Standards	4	2″ x 4″ x 5′	_____	_____
Ends	10	1″ x 6″ x 8′	_____	_____
Braces	4	1″ x 6″ x 6′	_____	_____
		Total Board Feet	_____	_____

Instructor's Score or Approval _____

> Job Sheet 18-5

Name _____ Date _____

Identifying How Materials Are Priced

Objective

When you have completed this activity you will have demonstrated the ability to identify how materials are priced.

Tools and Materials

Job sheet 18-5

Pencil or pen

Procedure

Complete the following worksheet.

___ 1. Lumber	a.	pound
___ 2. Sheet steel	b.	square foot or sheet
___ 3. Round steel	c.	standard length or foot
___ 4. Plywood	d.	piece or lineal foot or thousand board feet
___ 5. Pipe		
___ 6. Steel angles		

Note: An answer may be used more than once.

Instructor's Score or Approval _____

Unit 19 Selecting, Planning, and Building a Project

A good project will help you develop and measure your skills learned in agricultural mechanics. You may decide to use a plan, create one of your own, or modify an existing plan. Your decisions on projects should be discussed with your instructor.

Class Activity 19-1

Agricultural Mechanics Competency Inventory

Class Activity 19-2

Identifying Project Selection Criteria

Class Activity 19-3

Selecting a Project

Class Activity 19-4

Preparing a Plan

Class Activity 19-5

Identifying Design Principles

Job Sheet 19-1

Name _____ Date _____

Agricultural Mechanics Competency Inventory

Objective

When you have completed this activity you will have inventoried your agricultural mechanics competencies.

Tools and Materials

    Job sheet 19-1

    Pen or pencil

    Competency checklist

Procedure

Complete the competency checklist on the next page.

AGRICULTURAL MECHANICS COMPETENCY INVENTORY

INSTRUCTION: Check the column that best describes your personal level of experience in each area of competency.

	Level of Experience			
	1 Have Done	2 Am Familiar With	3 Will Learn This Year	4 No Skill; Will Not Learn This Year

1. Practice Safety and Shop Organization

a. Work safely with hand tools	☐	☐	☐	☐
b. Work safely with power tools	☐	☐	☐	☐
c. Follow safety rules in shop	☐	☐	☐	☐
d. Maintain safe work areas	☐	☐	☐	☐
e. Eliminate fire hazards	☐	☐	☐	☐
f. Select and use fire extinguishers	☐	☐	☐	☐
g. Interpret product labels	☐	☐	☐	☐
h. Act correctly in an emergency	☐	☐	☐	☐
i. Store materials correctly	☐	☐	☐	☐
j. Clean the shop effectively	☐	☐	☐	☐

2. Identify and Fit Agricultural Tools

a. Identify tools and equipment	☐	☐	☐	☐
b. Sharpen hand tools	☐	☐	☐	☐
c. Sharpen twist drills	☐	☐	☐	☐
d. Fit handles for hammers and axes	☐	☐	☐	☐
e. Fit handles for hoes and shovels	☐	☐	☐	☐
f. Dress grinding wheels	☐	☐	☐	☐

3. Maintain and Service a Home Shop

a. Clean and maintain an orderly shop	☐	☐	☐	☐
b. Observe safety regulations	☐	☐	☐	☐
c. Place and use fire-fighting equipment	☐	☐	☐	☐
d. Inventory tools and equipment	☐	☐	☐	☐
e. Plan shop layout	☐	☐	☐	☐
f. Select shop site	☐	☐	☐	☐

4. Apply Farm Carpentry Skills

a. Select lumber for a job	☐	☐	☐	☐
b. Operate power saws	☐	☐	☐	☐
c. Operate a jointer	☐	☐	☐	☐
d. Operate a planer	☐	☐	☐	☐
e. Saw to dimension	☐	☐	☐	☐
f. Bore holes	☐	☐	☐	☐
g. Glue wood	☐	☐	☐	☐
h. Drive and remove nails	☐	☐	☐	☐
i. Set screws and install bolts	☐	☐	☐	☐

Figure 19-1. Agricultural Mechanics Competencies

	Level of Experience			
	1 Have Done	2 Am Familiar With	3 Will Learn This Year	4 No Skill; Will Not Learn This Year

5. Properly Use Paint and Paint Equipment

a. Select paints or preservatives   ☐ ☐ ☐ ☐
b. Compute area for painting   ☐ ☐ ☐ ☐
c. Prepare wood surfaces   ☐ ☐ ☐ ☐
d. Prepare metal surfaces   ☐ ☐ ☐ ☐
e. Apply paint with a brush or roller   ☐ ☐ ☐ ☐
f. Clean and store paint brushes   ☐ ☐ ☐ ☐
g. Mask areas prior to painting   ☐ ☐ ☐ ☐
h. Apply paint with a spray can   ☐ ☐ ☐ ☐
i. Clean and store spray gun   ☐ ☐ ☐ ☐

6. Operate an Arc Welder

a. Practice safety in arc welding   ☐ ☐ ☐ ☐
b. Prepare metal for welding   ☐ ☐ ☐ ☐
c. Determine welder settings   ☐ ☐ ☐ ☐
d. Select electrodes   ☐ ☐ ☐ ☐
e. Operate AC and DC welders   ☐ ☐ ☐ ☐
f. Strike an arc   ☐ ☐ ☐ ☐
g. Run flat beads   ☐ ☐ ☐ ☐
h. Make weld joints   ☐ ☐ ☐ ☐
i. Weld horizontally   ☐ ☐ ☐ ☐
j. Weld vertically   ☐ ☐ ☐ ☐
k. Weld overhead   ☐ ☐ ☐ ☐
l. Cut with arc   ☐ ☐ ☐ ☐
m. Weld cast iron   ☐ ☐ ☐ ☐
n. Hard surface steel   ☐ ☐ ☐ ☐

7. Operate an Oxyacetylene Welder

a. Check for leaks   ☐ ☐ ☐ ☐
b. Turn equipment on and off   ☐ ☐ ☐ ☐
c. Adjust equipment   ☐ ☐ ☐ ☐
d. Change cylinders   ☐ ☐ ☐ ☐
e. Cut with cutting torch   ☐ ☐ ☐ ☐
f. Choose proper tips   ☐ ☐ ☐ ☐
g. Braze thin metal   ☐ ☐ ☐ ☐
h. Run beads   ☐ ☐ ☐ ☐
i. Make butt welds   ☐ ☐ ☐ ☐

8. Perform Skills in Hot and Cold Metal Work

a. Identify metals   ☐ ☐ ☐ ☐
b. Cut with hand hacksaw   ☐ ☐ ☐ ☐
c. Cut with cold chisel   ☐ ☐ ☐ ☐
d. Bend metal   ☐ ☐ ☐ ☐
e. Cut with tinsnips   ☐ ☐ ☐ ☐
f. Drill holes   ☐ ☐ ☐ ☐

Figure 19-1. Agricultural Mechanics Competencies (Continued)

	Level of Experience			
	1 Have Done	2 Am Familiar With	3 Will Learn This Year	4 No Skill; Will Not Learn This Year
g. Cut threads	☐	☐	☐	☐
h. Use files	☐	☐	☐	☐
i. Solder metal	☐	☐	☐	☐
j. Use power hacksaw	☐	☐	☐	☐
k. Heat metal with torch	☐	☐	☐	☐
l. Operate gas forge	☐	☐	☐	☐

9. Perform Skills in Concrete and Masonry Work

a. Build and prepare forms	☐	☐	☐	☐
b. Treat forms	☐	☐	☐	☐
c. Reinforce concrete	☐	☐	☐	☐
d. Test aggregates for impurities	☐	☐	☐	☐
e. Mix concrete	☐	☐	☐	☐
f. Pour concrete	☐	☐	☐	☐
g. Embed bolts	☐	☐	☐	☐
h. Finish concrete	☐	☐	☐	☐
i. Protect concrete while curing	☐	☐	☐	☐
j. Trowel concrete	☐	☐	☐	☐
k. Remove forms	☐	☐	☐	☐
l. Lay concrete blocks	☐	☐	☐	☐
m. Drill holes in concrete	☐	☐	☐	☐

10. Operate and Maintain Farm Engines

a. Read and follow operator's manual	☐	☐	☐	☐
b. Start and operate engines	☐	☐	☐	☐
c. Change oil and oil filters	☐	☐	☐	☐
d. Service air and fuel filters	☐	☐	☐	☐
e. Maintain battery water	☐	☐	☐	☐
f. Maintain and operate small gas engines	☐	☐	☐	☐
g. Identify engine components and systems	☐	☐	☐	☐
h. Remove and connect battery cables	☐	☐	☐	☐
i. Charge batteries	☐	☐	☐	☐
j. Read battery hydrometer	☐	☐	☐	☐
k. Test batteries	☐	☐	☐	☐
l. Troubleshoot fuel problems	☐	☐	☐	☐
m. Troubleshoot ignition problems	☐	☐	☐	☐

11. Perform Minor Tuneup and Repair of Farm Engines

a. Clean, gap, and replace spark plugs	☐	☐	☐	☐
b. Adjust carburetor mixture and speed crews	☐	☐	☐	☐
c. Install gap breaker points	☐	☐	☐	☐
d. Set dwell with meter	☐	☐	☐	☐
e. Install condensers	☐	☐	☐	☐
f. Time engines using timing light	☐	☐	☐	☐
g. Disassemble and reassemble distributors	☐	☐	☐	☐

Figure 19-1. Agricultural Mechanics Competencies (Continued)

	Level of Experience			
	1 Have Done	2 Am Familiar With	3 Will Learn This Year	4 No Skill; Will Not Learn This Year
h. Measure compression	☐	☐	☐	☐
i. Adjust valves	☐	☐	☐	☐
j. Set float levels in carburetors	☐	☐	☐	☐
k. Rebuild carburetors	☐	☐	☐	☐
l. Operate engine analyzers	☐	☐	☐	☐
m. Test and replace coils	☐	☐	☐	☐
n. Clean and install brushes	☐	☐	☐	☐
o. Clean commutator bars	☐	☐	☐	☐

12. Perform Electrification Skills

	1	2	3	4
a. Use safety measures in electrical wiring	☐	☐	☐	☐
b. Select correct fuse sizes	☐	☐	☐	☐
c. Replace fuses	☐	☐	☐	☐
d. Make splices	☐	☐	☐	☐
e. Repair electrical cords	☐	☐	☐	☐
f. Wire on-off switches	☐	☐	☐	☐
g. Select wire sizes	☐	☐	☐	☐
h. Know electrical terminology such as: volts, amps, watts, ohms	☐	☐	☐	☐
i. Attach wires to terminals	☐	☐	☐	☐
j. Solder splices	☐	☐	☐	☐
k. Install wire nut connectors	☐	☐	☐	☐
l. Install light fixtures	☐	☐	☐	☐
m. Install electric motors	☐	☐	☐	☐
n. Wire buildings and structures	☐	☐	☐	☐
o. Wire three-way switches	☐	☐	☐	☐
p. Wire four-way switches	☐	☐	☐	☐

13. Learn Plumbing Skills

	1	2	3	4
a. Repair leaky faucets	☐	☐	☐	☐
b. Assemble pipe and pipe fittings	☐	☐	☐	☐
c. Thread pipe	☐	☐	☐	☐
d. Measure and cut pipe	☐	☐	☐	☐
e. Cut plastic tubing/pipe	☐	☐	☐	☐
f. Install plastic tubing/pipe	☐	☐	☐	☐
g. Flare copper tubing	☐	☐	☐	☐
h. Cut copper tubing	☐	☐	☐	☐
i. Ream pipe	☐	☐	☐	☐
j. Install fixtures on plastic pipe	☐	☐	☐	☐
k. Sweat joints on copper pipe	☐	☐	☐	☐
l. Adjust air control mechanism on pressure system	☐	☐	☐	☐
m. Install pressure pumps	☐	☐	☐	☐
n. Repair pumps	☐	☐	☐	☐
o. Install water heaters	☐	☐	☐	☐
p. Cut soil pipe	☐	☐	☐	☐
q. Caulk joints on soil pipe	☐	☐	☐	☐
r. Lay soil pipe	☐	☐	☐	☐

Figure 19-1. Agricultural Mechanics Competencies (Continued)

Instructor's Score or Approval _____

Job Sheet 19-2

Name _____ Date _____

Identifying Project Selection Criteria

Objective

When you have completed this activity you will have demonstrated the ability to identify project selection criteria.

Tools and Materials

Job sheet 19-2

Pen or pencil

Procedure

Complete the following work sheet.

1. What is a project?

   _____

2. Why are projects used in the shop?

   _____
   _____
   _____

3. What are some of the skills a project might help you learn?

   _____
   _____
   _____
   _____

4. What are the recommended procedures for selecting a project?

   _____
   _____
   _____
   _____
   _____
   _____
   _____
   _____

Instructor's Score or Approval _____

Job Sheet 19-3

Name _____ Date _____

Selecting a Project

Objective

When you have completed this activity you will have demonstrated the ability to select a project.

Tools and Materials

    Job sheet 19-3

    Completed competency checklist

    Pen or pencil

    Plan books and ideas

Procedure

1. Discuss the competency checklist results with your instructor.
2. Ask your instructor for suggested projects to develop your skills.
3. Search the plan books for ideas.
4. Ask parents and others for ideas.
5. Look at sample projects.
6. Make a tentative decision.
7. Put away the plan books.

Instructor's Score or Approval _____

Job Sheet 19-4

Name _____ Date _____

Preparing a Plan

Objective

When you have completed this activity you will have demonstrated the ability to prepare a plan.

Tools and Materials

Job sheet 19-4

Graph paper with 1/4" squares

Pencil

Ruler

Procedure

1. Determine the longest dimension of the project.
2. Turn the paper horizontal for long projects and vertical for tall projects.
3. Divide 44 (squares on the paper) by the longest dimension of the project to find out how many feet or inches each square must represent.
4. Choose a scale that permits a one-view drawing with room for dimensions and notes.
5. Count the squares and draw the lines to complete one view with dimensions.
6. Do the other views on other sheets of paper.
7. Draw a plan as in Job Sheet 17-5.
8. Complete a bill of materials.
9. Clean the area and put away materials.

Instructor's Score or Approval _____

Job Sheet 19-5

Name _____ Date _____

Identifying Design Principles

Objective

When you have completed this activity you will have demonstrated the ability to identify basic design principles.

Tools and Materials

Job sheet 19-5

Pen or pencil

Procedure

Complete the following work sheet.

___ 1. Designs should be based on
___ 2. Use extra strength
___ 3. Structural steel
___ 4. Welding is stronger
___ 5. Secure fastening of wood
___ 6. Bolts are more secure than
___ 7. Screws are more secure than
___ 8. Crimps and folds
___ 9. Tongue and grooves or dowels and splines
___ 10. Wood and metal must be protected by

a. paint if exposed to weather
b. increase strength in wood joints
c. engineered plans
d. sheet metal panels
e. where safety is a factor
f. holds up better than wood
g. riveting
h. screws
i. glueing
j. nails

Instructor's Score or Approval _____

SECTION 6 TOOL FITTING

Unit 20 Repairing and Reconditioning Tools

Maintaining one's tools is the sign of a good workman. Properly sharpened and conditioned tools make work easier and more accurate and save time and energy in completing the task.

Learning how to care for tools is a valuable skill for the workman who takes pride in his work. Caring for these tools will also protect the financial investment in them.

Shop Activity 20-1

Restoring Leather Parts

Shop Activity 20-2

Restoring Wooden Parts

Shop Activity 20-3

Restoring Metal Surfaces

Shop Activity 20-4

Repairing Wooden Handles

Shop Activity 20-5

Replacing Wooden Handles

Shop Activity 20-6

Reshaping Tools

Shop Activity 20-7

Reshaping Toolheads

Job Sheet 20-1

Name _____ Date _____

Restoring Leather Parts

Objective

When you have completed this activity you will have demonstrated the ability to restore leather parts.

Tools and Materials

Job sheet 20-1

Leather to be restored

Saddle soap

Soft cloths

Sponge

Stiff bristle brush

Soft bristle brush

Water

Procedure

1. Brush off any dirt with the stiff bristle brush.
2. Dampen the sponge or cloth with water.
3. Rub the saddle soap until a lather forms.
4. Rub the lather onto all the leather surfaces until they are soft and clean.
5. Remove the lather with a cloth or sponge.
6. Allow the leather to dry.
7. Rub the leather with a soft cloth or brush.*
8. Clean the area and put away the materials.

  *Leather preservative oils may be applied at this time.

Instructor's Score or Approval _____

Job Sheet 20-2

Name _____ Date _____

Restoring Wooden Parts

Objective

When you have completed this activity you will have demonstrated the ability to restore wooden parts.

Tools and Materials

- Job sheet 20-2
- Wooden parts to be restored
- Fine sandpaper
- Soft cloths
- Stiff bristle brush
- Boiled linseed oil

Procedure

1. Brush away any dirt with the stiff bristle brush.
2. Sand the wooden parts to smooth rough areas.
3. Rub the wood with a soft cloth and linseed oil.
4. Repeat step 3 several times as needed.
5. Allow the wood to dry.
6. Buff with the soft, dry cloth.
7. Clean the area and put away the materials.

Note: Wax can be used to do the same job using the same procedures.

Instructor's Score or Approval _____

Job Sheet 20-3

Name _____ Date _____

Restoring Metal Surfaces

Objective

When you have completed this activity you will have demonstrated the ability to restore metal surfaces.

Tools and Materials

    Job sheet 20-3

    Metal to be restored

    Wire brush

    Putty knife

    Commercial solvent

    Fine sandpaper

    Light weight oil

    Paper towels

    Soft cloths

Procedure

1. Tap the metal to remove hardened materials.
2. Use the putty knife to remove hardened materials.
3. Use the wire brush to removed pitted rust.
4. Use 400 sandpaper to restore a smooth surface.
5. Use the commercial solvent to clean the tool.
6. Dry the tool with towels or cloths.
7. Apply a light coat of oil.
8. Clean the area and put away the materials.

Instructor's Score or Approval _____

Job Sheet 20-4

Name _____ Date _____

Repairing Wooden Handles

Objective

When you have completed this activity you will have demonstrated the ability to repair a wooden handle.

Tools and Materials

    Job sheet 20-4

    Handle to be repaired

    Wood glue

    Sandpaper

    Cloths

    Boiled linseed oil

    Vise

    Clamps

Procedure

1. Clamp the tool.
2. Force the split open and glue both sides.
3. Clamp the handle in several places.
4. Clean off excess glue.
5. Allow to dry.
6. Remove the clamps and sand the handle.
7. Treat the handle with linseed oil.
8. Polish when dry.
9. Clean the area and put away the materials.

Instructor's Score or Approval _____

> **Job Sheet 20-5**

Name _____ Date _____

Replacing Wooden Handles

Objective

When you have completed this activity you will have demonstrated the ability to replace a wooden handle.

Tools and Materials

Job sheet 20-5	Boiled linseed oil
Tool needing the handle	Plastic container the size of the tool
Tool handle	Wood rasp
Vise	Wooden mallet
Wooden wedge	Portable drill and bits
Metal wedges	Hacksaw
Punch	Handsaw

Procedure

1. Clamp the toolhead firmly in the vise.
2. Drill the wooden remains of the old handle.
3. Drive out the remainder of the old handle.
4. Check the new handle to see where wood needs to be removed.
5. Rasp the wood to a trial fit.
6. When the fit is snug, mark the handle on both sides of the head.
7. Saw a kerf along the long center line of the handle 2/3 of the depth of the head.
8. Reposition the handle and squeeze the handle until the kerf is closed.
9. Resaw the kerf.
10. Drive the handle securely with the mallet.
11. Grip the handle in the vise just below the head.
12. Saw the handle flush with the hacksaw.
13. Drive in the wooden wedge until it spreads the handle and fills the head.
14. Saw the wooden wedge flush.
15. Drive one or two steel wedges across the wooden wedge to hold it in place.
16. Place the tool in a container of boiled linseed oil for several days.
17. Dry the tool and treat the handle with linseed oil.
18. Allow the handle to dry.
19. Buff the handle with the soft cloth.
20. Clean the area and put away the materials.

Instructor's Score or Approval _____

Job Sheet 20-6

Name _____ Date _____

Reshaping Tools

Objective

When you have completed this activity you will have demonstrated the ability to reshape a tool.

Tools and Materials

    Job sheet 20-6

    Tool to be reshaped

    File

    Grinding wheel

    Face shield

    Leather gloves

    Container of water

    Protective clothing

Procedure

1. Examine a similar tool to guide your shaping.
2. Put on the face shield.
3. Put on the protective clothing.
4. Put on the leather gloves.
5. Check the face of the grinding wheel to see that it is flat.
6. Adjust the tool rest parallel to the shaft.
7. For a screwdriver or punch, square the face of the tool on the grinding wheel.
8. Taper the sides to reestablish the fit of the screwdriver in the slot on a screw.
9. Avoid excessive heat in grinding using water to cool the tool.
10. Dress the sides of the screwdriver blade to complete the shaping.
11. Clean the area and put away the materials.

Instructor's Score or Approval _____

Job Sheet 20-7

Name _____ Date _____

Reshaping Toolheads

Objective

When you have completed this activity you will have demonstrated the ability to reshape a toolhead.

Tools and Materials

Job sheet 20-7

Tool to be reshaped

Face shield

Protective clothing

Leather gloves

Grinding wheel

Container of water

Procedure

1. Examine a similar tool to guide your shaping.
2. Put on the face shield.
3. Put on the protective clothing.
4. Put on the leather gloves.
5. Check the grinding wheel for chips or nicks.
6. Adjust the tool rest parallel to the shaft.
7. Slowly grind a taper from each flat surface to the crown.
8. Avoid overheating the tool by using water.
9. Finish by twirling the tool to provide a rounded crown.
10. Clean the area and put away the materials.

Instructor's Score or Approval _____

Unit 21 Sharpening Tools

Sharp tools mark the good craftsman. Sharp tools require less energy, do finer work and are safer to work with for the worker. Sharpening tools are used for sharpening other tools. Learning how to sharpen tools properly will enable you to do better quality work faster. Knowing the proper angles for cutting edges is important to obtain a properly working tool. Keeping a copy of the information in text Figure 21-3 available will help you keep your tools in proper working condition. Cutting tools use the simple lever principle of the inclined plane; some use a single inclined plane and others a double inclined plane.

Read Unit 21 and refer to it as needed as you complete the following activities.

Class Activity 21-1

Identifying Sharpening Tools and Principles

Shop Activity 21-2

Making a Tool-Sharpening Gauge

Shop Activity 21-3

Sharpening a Center Punch

Shop Activity 21-4

Sharpening a Wood Chisel

Shop Activity 21-5

Sharpening a Knife

Shop Activity 21-6

Sharpening a Cold Chisel

Shop Activity 21-7

Sharpening an Axe or Hatchet

Shop Activity 21-8

Sharpening a Twist Bit

Shop Activity 21-9

Sharpening a Rotary Mower Blade

Shop Activity 21-10

Sharpening a Digging Tool

WARNING: Sharpening activities generate heat. Care must be observed to control the heat so that the tool temper is preserved.

WARNING: Eye protection is most important in sharpening activities to protect from flying particles.

Job Sheet 21-1

Name _____ Date _____

Identifying Sharpening Tools and Principles

Objective

Upon completion of this activity you will have demonstrated the ability to identify sharpening tools and principles.

Tools and Materials

Job sheet 21-1

Pen or pencil

Procedure

Using the text as a reference, complete the following worksheet.

1. Why are sharp tools safer to use? _____
2. What are the tools used to sharpen other tools? _____
3. T F Cutting tools usually have either a single or a double cutting edge.
4. Check which tools have single and which have double inclined planes as cutting edges.

Tool	Single	Double
Wood chisels	_____	_____
Cold chisels	_____	_____
Knives	_____	_____
Axes	_____	_____
Shears	_____	_____
Hatchets	_____	_____
Drills	_____	_____
Digging tools	_____	_____
Saws	_____	_____

5. Tools with a single inclined plane are sharpened from _____ side(s).
6. Tools with double inclined planes are sharpened from _____ side(s).
7. T F Edges that are convex require more force for cutting.
8. T F Edges that are concave require more force for cutting.
9. Saws with teeth wider than the blade are said to be _____
10. Describe a simple test for tool sharpness. _____
11. What is temper? _____
12. Why is it important? _____
13. What is whetting? When is it used? _____
    _____

Instructor's Score or Approval _____

Job Sheet 21-2

Name _____ Date _____

Making a Tool-Sharpening Gauge

Objective

Upon completing this activity you will have demonstrated the ability to make a useful tool sharpening gauge.

Tools and Materials

Job sheet 21-2	Bench rule	Drill press
26 gauge galvanized stock	Dividers	1/4" twist drill
Tinsnips	Flat file	Safety glasses or face shield
Hacksaw	Center punch	
Scribe	Protractor	

Procedure

1. Cut a 3" x 4 1/2" piece of stock with all corners square.
2. Measure in from one end 1 1/2" and mark the location.
3. Square a line across the stock through the marked location.
4. Measure along this line 1 1/2" and lightly mark this location with the center punch.
5. Set the dividers at 1 1/2".
6. Place one leg of the dividers in the punch mark and scribe a semicircle across the end of the stock.
7. Set the dividers to 1 3/16".
8. Scribe a second semicircle inside the first arc.
9. Center the protractor on the scribed line and mark the inner semicircle at 20, 60, 120, and 160 degrees.
10. Using the protractor, lay out and cut triangles of 20, 45, 75, and 80 degrees from heavy paper to use as patterns.
11. Place the point of the pattern for the 20 degree cutout on the 20 degree mark on the inner semicircle and scribe the pattern.
12. Place the point of the pattern for the 45 degree cutout at the 60 degree mark on the inner semicircle and scribe the pattern.
13. Place the point of the pattern for the 75 degree cutout on the 120 degree mark on the inner semicircle and scribe the pattern.
14. Place the point of the 80 degree pattern on the 160 degree mark of the inner semicircle and scribe the pattern.
15. Cut the waste from the outer semicircle with the tinsnips.
16. Cut out the four scribed gauges with the tinsnips.
17. Measure 5/16" along the scribed line from the left side of the stock and mark the location.
18. Repeat at the square end of the stock.
19. Scribe a line between these two locations.
20. Measure 9/16" from the scribed line toward the square end of the stock and mark the location (along the left side of the stock).
21. Scribe a line from this point to the point on the first scribed line.

22. Cut out the 29 degree gauge just scribed.
23. Measure along the first scribed line 3/4″ from the right side and mark the location.
24. Repeat at the square end.
25. Scribe a line between these two locations.
26. Measure 13/32″ from the first scribed line along the newly scribed line and mark this location.
27. Scribe a line from this location to the right end of the first scribed line.
28. Cut out this 118 degree gauge.
29. Measure in 3/8″ from the square end and mark the location.
30. Square through this location and mark the center with the center punch.
31. Drill a 1/4″ hole at this location.
32. Using the file, smooth all corners and edges.
33. Clean area and store tools.

Instructor's Score or Approval _____

Lab Manual

Job Sheet 21-3

Name _____ Date _____

Sharpening a Center Punch

Objective

Upon the completion of this activity you will have demonstrated the ability to sharpen a center punch.

Tools and Materials

Job sheet 21-3

Grinding wheel

Tool gauge

Center punch needing sharpening

Safety glasses or a face shield

Procedure

1. Check the angle of the center punch with the 75 degree tool gauge.
2. Set the tool rest on the grinding wheel to support the punch at the correct angle.
3. Use the fingers of the left hand to hold the punch against the tool rest as a guide and twirl the punch with the right hand to obtain a sharp point.
4. Check the angle of the point with the 75 degree tool gauge.
5. Repeat steps 4 and 5 until the proper point is obtained.
6. Clean area and store tools.

Instructor's Score or Approval _____

Job Sheet 21-4

Name _____ Date _____

Sharpening a Wood Chisel

Objective

Upon completing this activity you will have demonstrated the ability to sharpen a wood chisel.

Tools and Materials

Job sheet 21-4

Wood chisel in need of sharpening

Grinding wheel

Oil stone & cutting oil

Tool sharpening gauge

Try or combination square

Safety glasses or a face shield

Procedure

1. Check the angle of the wood chisel with the 29 degree tool gauge.
2. Set the tool rest on the grinding wheel to support the chisel at the correct angle.
3. Hold the chisel with the left hand acting as the guide against the tool rest and move the chisel smoothly across the wheel with the right hand to obtain a smooth grind.
4. Check the squareness of the blade with the square.
5. Repeat steps 3 and 4 until the proper square angle is obtained.
6. Hone the chisel by placing the back of the chisel flat on the oilstone and drawing it back and forth to remove the wire edge or curl.
7. Hone the face of the chisel by placing the ground edge flat on the stone and raising the heel slightly and moving the chisel in a figure 8 motion over the stone using oil as needed.
8. Test the sharpness of the chisel.
9. Repeat steps 6 to 8 until the desired sharpness is achieved.
10. Wipe the slurry from the oilstone and store the tools.

Caution: Don't overheat the chisel while grinding; use cold water to cool the chisel if necessary.

Instructor's Score or Approval _____

Job Sheet 21-5

Name _____ Date _____

Sharpening a Knife

Objective

Upon completing this activity you will have demonstrated the ability to sharpen a knife.

Tools and Materials

Job sheet 21-5

Knife in need of sharpening

Grinding wheel

Oilstone and cutting oil

Wiping cloth

Safety glasses or a face shield

Procedure

1. Check the cutting angle of the knife with the 29 degree gauge.
2. Remove any nicks by grinding.
3. Hold the knife with the cutting edge up and move it back and forth across the grinding wheel to create an even grind mark of about 1/4" from the edge of the blade.
4. Use a fine grit wheel to remove the grinding marks if possible.
5. Whet the knife on the oilstone by placing the knife flat on the stone and raising the back slightly, alternately drawing the blade toward the cutting edge then flipping the knife over and pushing it in the other direction.
6. Test the blade's sharpness.
7. Repeat steps 5 and 6 until the correct degree of sharpness is obtained.
8. Wipe the oilstone and store the tools.

Caution: Grind slowly to avoid overheating the edge; use cold water to cool the blade if necessary.

Instructor's Score or Approval _____

Job Sheet 21-6

Name _____ Date _____

Sharpening a Cold Chisel

Objective

Upon completing this activity you will have demonstrated the ability to sharpen a cold chisel.

Tools and Materials

Job sheet 21-6

Cold chisel

Grinding wheel

Tool grinding gauge

Try or combination square

Safety glasses or a face shield

Procedure

1. Check the angle of the cutting edge.
2. Hold the chisel on the tool rest with the fingers as the guide and move the chisel across the wheel, then rotate the chisel one half turn and repeat.
3. Check the cutting angle and squareness.
4. Repeat steps 2 and 3 until a sharp cutting edge is obtained.
5. Clean area and store the tools.

Caution: Avoid overheating the chisel.

Instructor's Score or Approval _____

Job Sheet 21-7

Name _____ Date _____

Sharpening an Axe or Hatchet

Objective

Upon completing this activity you will have demonstrated the ability to sharpen an axe or hatchet.

Tools and Materials

    Job sheet 21-7

    Axe or hatchet needing sharpening

    Flat file or hand stone

    Grinding wheel or portable grinder

    Safety glasses or a face shield

    Leather gloves

Procedure

1. Check the blade angle with the 20 degree gauge.
2. Remove any nicks by grinding.
3. Draw a line parallel and 1/2" to 5/8" from the edge of the blade.
4. Hold the blade so it forms a 10 degree angle with the wheel and move it up and down and across the wheel with your gloved hands to create the wide grinding band and preserve the cutting angle.
5. Reverse the blade and repeat the grinding.
6. Repeat steps 4 and 5 until a good edge is obtained.
7. Remove the grinding marks and wire edge with the flat file or hand stone.
8. If a portable grinder is used, secure the blade in a machinist vise.
9. Hold the portable grinder at the appropriate 10 degree angle and grind the blade to retain the proper angle.
10. Reverse the blade and repeat step 9.
11. Repeat steps 9 and 10 until a proper edge is obtained.
12. Use the hand stone or flat file to complete the edge.
13. Clean area and store the tools.

Caution: The grinding creates heat; cool the blade as necessary by cold water to preserve the temper.

Instructor's Score or Approval _____

Job Sheet 21-8

Name _____ Date _____

Sharpening a Twist Bit

Objective

Upon the completion of this activity you will have demonstrated the ability to sharpen a twist bit.

Tools and Materials

Job sheet 21-8

Twist bit in need of sharpening

Grinding wheel

Tool grinding gauge

Safety glasses or a face shield

Procedure

1. Check the bit to determine:
   a. if the lip fits the 118 degree gauge.
   b. if the heel has 10 to 12 degrees of clearance.
   c. if the dead center is the center of the bit.
2. Check the face of the grinding wheel for straightness and square to the edge and dress as needed.
3. Set the tool rest horizontal to the center of the wheel.
4. Hold the drill between your thumb and index finger with an inch exposed and with the face parallel to the face of the wheel.
5. With the right hand holding the end of the bit, touch the lip to the wheel and lower the end of the bit while giving it a slight clockwise twist.
6. Turn the bit a half turn and repeat.
7. Repeat step 1.
8. Repeat steps 5 to 7 as needed.
9. Test the bit by drilling mild steel. An off-center drill will not drill an accurately sized hole.
10. Clean area and store tools.

Caution: Avoid overheating the bit; use cold water as needed to cool the bit.

Instructor's Score or Approval _____

Job Sheet 21-9

Name _____ Date _____

Sharpening a Rotary Mower Blade

Objective

Upon completing this activity you will have demonstrated the ability to sharpen a rotary mower blade.

Tools and Materials

Job sheet 21-9

Tool grinding gauge

Portable grinder

Blade in need of sharpening

Flat file or grinding wheel

Blade balancer

Safety glasses or a face shield

Procedure

1. Check the cutting angle of the blade with the 45 degree gauge.
2. Remove any nicks by filing or grinding.
3. Secure the blade in a vise.
4. Restore the flat side of the cutting edge with the portable grinder.
5. Reverse the blade and repeat step 4.
6. Test the blade cutting angle with the 45 degree gauge.
7. Test the blade balance using a nail as a pivot if no balancer is available, and grind more on the heavy side as needed.
8. Repeat steps 4 to 7 as necessary.
9. Clean area and store the tools.

Caution: Avoid overheating the blade; use cold water to cool it as necessary.

Instructor's Score or Approval _____

Job Sheet 21-10

Name _____ Date _____

Sharpening a Digging Tool

Objective

Upon completing this activity you will have demonstrated the ability to sharpen a digging tool.

Tools and Materials

    Job sheet 21-10

    Hand stone

    Hoe, shovel, or spade needing sharpening

    Flat file

    Machinist's vise

    Safety glasses or a face shield

Procedure

1. Check the original cutting angle.
2. Secure the tool in the vise.
3. Straighten or remove any bends or curls.
4. Restore the cutting edge with the file.
5. Check the original cutting angle with the gauge as in step 1.
6. Repeat steps 4 and 5 as necessary.
7. Clean the area and put away the tools.

Instructor's Score or Approval _____

SECTION 7 GAS HEATING, CUTTING, BRAZING, AND WELDING

Unit 22 Using Gas Welding Equipment

Gases that burn offer us many advantages and at the same time expose us to the danger of fire or explosion. Safety is of primary importance when dealing with compressed and flammable gases. This author lived in a town where a mechanic accidentally tipped over the oxygen tank and the regulator broke off. The tank went through two concrete block walls and down an alley; fortunately no one was injured.

Class Activity 22-1

Identifying Safety Rules with Compressed Gases

Class Activity 22-2

Understanding the Use of Compressed Gases

Shop Activity 22-3

Setting Up the Oxyacetylene Torch

Shop Activity 22-4

Pressurizing the Torch

Shop Activity 22-5

Lighting, Adjusting, and Shutting Down the Torch

Class Activity 22-6

Identifying Oxyacetylene Flames

Job Sheet 22-1

Name _____ Date _____

Identifying Safety Rules with Compressed Gases

Objective

When you have completed this activity you will have demonstrated the ability to identify safety rules for compressed gases.

Tools and Materials

    Job sheet 22-1

    Pen or pencil

Procedure

    Complete the following worksheet.

___ 1.	Wear a face shield	a.	spontaneous fire possibility
___ 2.	Before using gases	b.	locked chain link fence or block buildings
___ 3.	Store separately	c.	wear leather gloves and apron
___ 4.	Chained securely	d.	any cylinder not having regulators attached
___ 5.	Store cylinders	e.	at all times
___ 6.	Keep caps on	f.	obtain permission
___ 7.	Follow specific procedures	g.	approved fire extinguisher in the area
___ 8.	Never expose equipment to oil or grease	h.	check thoroughly for leaks
		i.	for turning systems on or off
___ 9.	Always wear when using gas equipment	j.	point away from everyone or clothing
___ 10.	Learn to recognize	k.	turn off the gas at the tank
___ 11.	Protect gas cylinder storage areas	l.	the odors of combustible fuels
		m.	it can be saturated with oxygen or fuel gases
___ 12.	In case the equipment catches on fire	n.	fuel and oxygen cylinders
		o.	free from materials that will burn
___ 13.	All equipment or cylinders that can discharge gas	p.	equipment connected to compressed gas cylinders
___ 14.	Never leave clothing where	q.	in an upright position
___ 15.	When connections are opened or cylinders changed	r.	outdoors or in a well-ventilated area
___ 16.	Work only in areas that are		
___ 17.	Do not bump or put pressure on		
___ 18.	Never work in an area without an		

Instructor's Score or Approval _____

Job Sheet 22-2

Name _____ Date _____

Understanding the Use of Compressed Gases

Objective

   Upon completion of this activity you will have demonstrated an understanding of the use of compressed gases.

Tools and Materials

   Job sheet 22-2

   Pen or pencil

Procedure

1. Define fusion.
   _____
   _____

2. Define welding.
   _____
   _____

3. What are the dangers of compressed gases?
   _____
   _____

4. What does flammable mean?
   _____

5. What are the most common gases for heating, cutting, and welding?
   _____

6. What is a manifold? Why are they used?
   _____

7. What is oxyacetylene?
   _____
   _____

8. Acetylene hoses are colored _____. Oxygen hoses are colored _____.

9. The acetylene wrench must be left on the cylinder at all times when in use because
   _____

10. Why are the threads on oxygen and acetylene hoses and fittings different?
    _____

11. How should the equipment be tested for leaks?
    _____

12. Lens shade number _____ is recommended for oxyacetylene use.

Instructor's Score or Approval _____

Job Sheet 22-3

Name _____ Date _____

Setting Up the Oxyacetylene Torch

Objective

When you have completed this activity you will have demonstrated the ability to set up the oxyacetylene torch.

Tools and Materials

    Job sheet 22-3

    Acetylene regulator

    Oxygen regulator

    Acetylene hose

    Oxygen hose

    Oxygen tank

    Acetylene tank

    Torch and tips

    Face shield

    Adjustable wrench

    Oxyacetylene cart

Procedure

1. Obtain permission.
2. Put on the face shield.
3. Chain the two tanks to the cart.
4. Open and close the acetylene tank valve quickly to blow out any dirt.
5. Attach the acetylene regulator (left-handed threads) to the acetylene tank.
6. Attach the red hose to the acetylene gauge outlet.
7. Attach the other end of the red hose to the torch body.
8. Remove the protective cap from the oxygen tank.
9. Crack the oxygen cylinder to blow out any dirt.
10. Attach the oxygen regulator to the oxygen tank.
11. Attach the green hose to the oxygen regulator gauge outlet.
12. Attach the other end of the green hose to the oxygen inlet on the torch body.
13. Attach the torch tip to the torch body.
14. Clean the area and put away the materials.

Instructor's Score or Approval _____

Job Sheet 22-4

Name _____ Date _____

Pressurizing the Torch

Objective

When you have completed this activity you will have demonstrated the ability to pressurize the torch.

Tools and Materials

    Job sheet 22-4

    Oxyacetylene equipment

    Face shield

    Leather gloves

    Leather apron

Procedure

1. Obtain permission.
2. Put on the face shield.
3. Put on the leather apron.
4. Put on the leather gloves.
5. Close the acetylene valve on the torch.
6. Close the oxygen valve on the torch.
7. Turn the acetylene regulator handle counterclockwise until no tension is felt.
8. Turn the oxygen regulator handle counterclockwise until no tension is felt.
9. Open the oxygen cylinder valve slowly until the regulator gauge responds, then open it fully.
10. Open the acetylene cylinder gauge slowly 1/2 turn.
11. Open the oxygen torch valve 1/8 turn and the regulator handle clockwise until the gauge reads 10 psi.
12. Close the oxygen torch valve.
13. Open the acetylene torch valve 1/8 turn and turn the acetylene regulator valve clockwise until the gauge reads 5 psi.
14. Close the acetylene torch valve.
15. If this is an initial setup, check for leaks with soapy water.

Instructor's Score or Approval _____

Job Sheet 22-5

Name _____ Date _____

Lighting, Adjusting, and Shutting Down the Torch

Objective

When you have completed this activity you will have demonstrated the ability to light, adjust, and shut down the torch.

Tools and Materials

    Job sheet 22-5

    Oxyacetylene equipment

    Face shield

    Leather gloves

    Leather apron

    Spark lighter

    Shade 5 welding goggles

Procedure

1. Obtain permission.
2. Put on the shade 5 welding goggles.
3. Put on the face shield.
4. Put on the leather apron.
5. Put on the leather gloves.
6. Pressurize the torch.
7. Open the acetylene valve 1/8 turn.
8. Ignite the torch with the spark lighter.
9. Open the acetylene valve slowly until the flame is 1/4" off the tip.
10. Open the oxygen valve slowly until the long inner flame matches the cone giving a neutral flame.
11. Use more oxygen only when special circumstances require.
12. Turn off the acetylene at the cylinder.
13. Turn off the oxygen at the cylinder.
14. Open the torch valves until the regulators both return to zero.
15. Close the torch valves.
16. Turn the regulator handles counterclockwise until no tension is felt.
17. Coil the hoses and store the equipment.

Instructor's Score or Approval _____

Job Sheet 22-6

Name _____ Date _____

Identifying Oxyacetylene Flames

Objective

Upon completing this activity you will have demonstrated the ability to identify carbonizing, neutral, and oxidizing flames.

Tools and Materials

   Job sheet 22-6

   Pen or pencil

Procedure

Complete the following work sheet by identifying each type of flame.

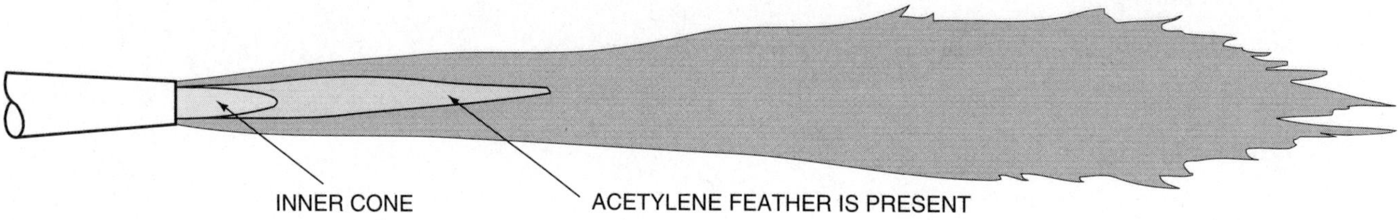

INNER CONE   ACETYLENE FEATHER IS PRESENT

_____

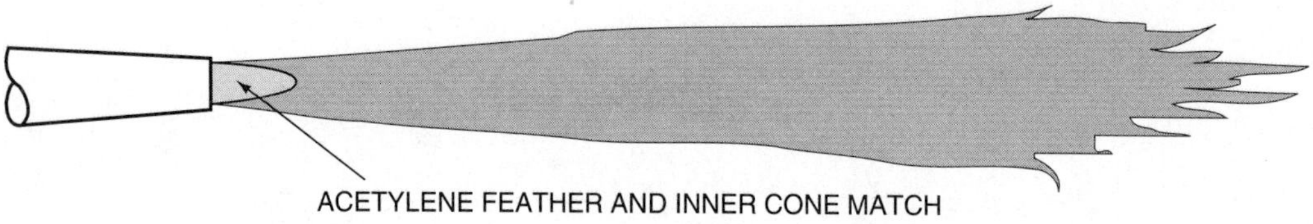

ACETYLENE FEATHER AND INNER CONE MATCH

_____

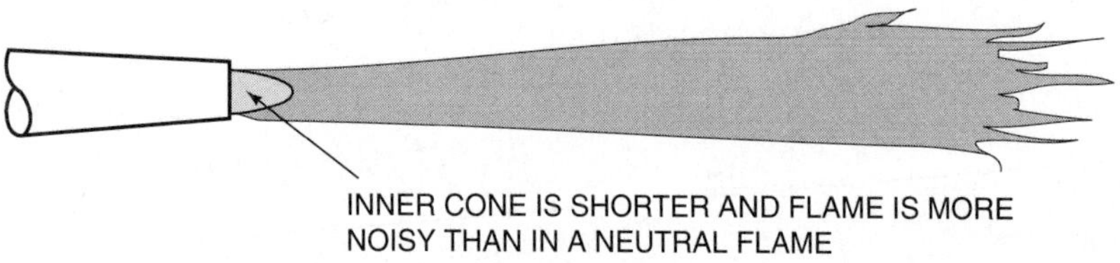

INNER CONE IS SHORTER AND FLAME IS MORE NOISY THAN IN A NEUTRAL FLAME

_____

Figure 22-1. Oxyacetylene Flames

Instructor's Score or Approval _____

Unit 23 Cutting with Oxyfuels and Other Gases

Oxyfuels refer to a combination of oxygen and a fuel gas such as acetylene, propane, natural gas, or MAPP. These, in combination with oxygen, burn with intense heat making possible cutting, brazing, and welding activities.

Class Activity 23-1

Understanding the Use of Oxyfuels

Shop Activity 23-2

Cutting Steel With Oxyfuels

Lab Manual

Job Sheet 23-1

Name _____ Date _____

Understanding the Use of Oxyfuels and Other Gases

Objective

Upon completing this activity you will have demonstrated an understanding of the use of oxyfuels.

Tools and Materials

Job sheet 23-1   Pen or pencil

Procedure

Complete the following worksheet.

1. Define the term oxyfuels. _____
   _____

2. Burning oxyfuels may reach a temperature of _____ F.

3. Burning oxyfuels are used for _____, _____,
   _____, and _____.

4. Which is the most common fuel for welding? _____

5. Why are other gases preferred?
   _____

6. Acetylene pressure is recommended not to exceed _____ psi.

7. What is MAPP gas? _____
   _____

8. How does one correct for too much preheating when cutting?
   _____

9. How do you identify the problem?
   _____

10. What is the proper torch clearance from the stock?
    _____
    _____

11. How do you identify proper speed as compared to fast or slow travel?
    _____
    _____

12. How do you identify pressure problems?
    _____
    _____

13. What is plasma arc cutting?
    _____
    _____

Instructor's Score or Approval _____

Job Sheet 23-2

Name _____ Date _____

Cutting Steel With Oxyfuels

Objective

When you have completed this activity you will have demonstrated the ability to cut steel with oxyfuels.

Tools and Materials

Job sheet 23-2   Leather gloves

Oxyfuel equipment   Leather apron

Cutting torch   Steel to be cut

Shade 5 goggles   Pail of sand

Face shield

Procedure

1. Obtain permission.
2. Put on the shade 5 goggles.
3. Put on the face shield
4. Put on the leather apron.
5. Put on the leather gloves.
6. Clear the area of combustibles.
7. Place the sand to catch the hot metal.
8. Select the proper tip for the job.
9. Set the pressures for the tip used (see text Figure 23-5).
10. Light the torch using the acetylene gas only.
11. Open the valve until the flame leaves the tip then close it until the flame touches the tip.
12. With the oxygen preheat valve closed, open the oxygen valve several turns.
13. Slowly open the preheat valve until the acetylene feathers just match the inner cone of the preheat flame.
14. Press the oxygen cutting lever and observe whether the flames stay neutral.
15. Adjust to a neutral flame.
16. Hold the flame over the corner of the cut to be made.
17. Hold the inner cone about 1/8" above the metal until it turns cherry red.
18. Press the cutting lever and move slowly through the metal.
19. Shut down the torch and purge the lines.
20. Allow the metal to cool.
21. Clean the area and put away the materials.

Instructor's Score or Approval _____

Unit 24 Brazing and Welding with Oxyacetylene

Brazing is the bonding of metals by alloys melting at temperatures above 840 degrees. Soldering is the bonding of metals with alloys that melt below 840 degrees. Learning these skills will be valuable additions to your knowledge and skills.

Class Activity 24-1

Understanding Brazing and Welding with Oxyacetylene

Shop Activity 24-2

Using Flux

Shop Activity 24-3

Tinning with Solder

Shop Activity 24-4

Tinning with Braze Filler

Shop Activity 24-5

Running a Bead

Shop Activity 24-6

Brazing

Shop Activity 24-7

Braze Welding Butt Welds

Shop Activity 24-8

Braze Welding Fillet Welds

Shop Activity 24-9

Pushing the Puddle

Shop Activity 24-10

Making a Corner Weld Without Filler Rod

Job Sheet 24-1

Name _____ Date _____

Understanding Brazing and Welding with Oxyacetylene

Objective

Upon completing this activity you will have demonstrated an understanding of brazing and welding with oxyacetylene.

Tools and Materials

Job sheet 24-1

Pen or pencil

Procedure

Complete the following worksheet.

1. What do brazing, soldering, and fusion welding have in common?
   _____
   _____

2. Define brazing. _____
   _____

3. What is capillary action? _____
   _____

4. How does capillary action aid in soldering and brazing? _____
   _____

5. What is chemical cleaning? _____
   _____

6. What is a flux? _____
   _____

7. Why should care be taken to avoid overheating? _____
   _____

8. How can we successfully heat two different thicknesses of metals at the same time?
   _____
   _____

9. What is the general rule for torch adjustment? _____
   _____

10. How do you prevent burn-through when welding? _____
    _____

11. What do we mean by pushing the puddle? _____
    _____
    _____

Instructor's Score or Approval _____

Job Sheet 24-2

Name _____ Date _____

Using Flux

Objective

When you have completed this activity you will have demonstrated the ability to use flux.

Tools and Materials

Job sheet 24-2

Copper pipe, tubing, or sheet

Rosin flux

Shade 5 welding goggles

Face shield

Leather gloves

Steel wool or sandpaper

Leather apron

Oxyacetylene welding outfit

Procedure

1. Obtain permission.
2. Put on the protective clothing.
3. Set up, light, and adjust a small torch.
4. Put a small amount of rosin flux on the copper.
5. Heat the area by moving the flame in and out of the area of the flux.
6. Remove the heat when the copper suddenly becomes shiny.
7. Move the flame to an area adjacent and watch the flux flow to the new area.
8. Continue to heat the copper until the heat burns the flux and blackens the area.
9. Remove the burned flux with sandpaper or steel wool.
10. Reapply the flux and reheat.*
11. Clean the area and put away the materials.

    *Note that the fluxing action can only be repeated after the mechanical cleaning of the area. The ability to control the fluxing action is necessary for successful bonding of solder or brazing materials.

Instructor's Score or Approval _____

Job Sheet 24-3

Name _____ Date _____

Tinning with Solder

Objective

When you have completed this activity you will have demonstrated the ability to tin with solder.

Tools and Materials

    Job sheet 24-3

    A piece of copper

    Oxyfuel torch

    Shade 5 welding goggles

    Face shield

    Leather gloves

    Leather apron

    Solder

    Fine emery cloth

    Paste flux

    Damp cloth

Procedure

1. Obtain permission.
2. Put on the protective clothing.
3. Clean the copper with the emery cloth.
4. Light and adjust an oxyfuel or propane torch.
5. Heat the copper until the fluxing action starts.
6. Touch the solder to the metal and feed it down as the copper is tinned.
7. Move the heat until the area is tinned.
8. Smooth and even the solder with the flame.
9. Wipe the area quickly with the damp cloth to remove excess solder and improve appearance.
10. Clean the area and put away the materials.

Instructor's Score or Approval _____

> **Job Sheet 24-4**

Name _____ Date _____

Tinning with Braze Filler

Objective

When you have completed this activity you will have demonstrated the ability to tin with braze filler.

Tools and Materials

    Job sheet 24-4

    Oxyfuel outfit

    Braze filler rod

    Flux

    Face shield

    Leather gloves

    Leather apron

    Shade 5 welding goggles

    Fine emery cloth

    Firebricks

    Nonrusted steel plate approximately 1/8" x 2" x 6"

Procedure

1. Obtain permission.
2. Put on the protective clothing.
3. Clean the steel with the emery cloth.
4. Place the steel on the firebricks.
5. Light and adjust the torch using a small tip.
6. Heat the steel until the center is dull red.
7. As you heat the base metal, heat the filler rod by holding it in the flame.
8. Flux the heated rod by dipping it in the flux can.
9. Place the flux-coated rod in the flame and touch the base metal.
10. When the temperature is correct, fluxing action takes place, cleaning impurities.
11. As fluxing action occurs, watch for the filler rod to melt, playing the flame over the metal, the rod, and flux.
12. Control the temperature to tin the whole area.
13. Clean the area and put away the materials.

Instructor's Score or Approval _____

Job Sheet 24-5

Name _____ Date _____

Running a Bead

Objective

When you have completed this activity you will have demonstrated the ability to run a bead.

Tools and Materials

Job sheet 24-5

Oxyfuel torch

Shade 5 welding goggles

Face shield

Leather gloves

Leather apron

Clean 1/8" x 2" x 6" steel

Flux

Firebricks

Emery cloth

Filler rod

Procedure

1. Obtain permission.
2. Put on the protective clothing.
3. Clean the metal with the emery cloth.
4. Put the metal on a firebrick.
5. Light and adjust the torch.
6. Heat the metal 1/2" from the end (run the bead across the metal).
7. Heat the rod and dip it in the flux.
8. Touch the fluxed rod to the metal and, as it melts, move it in and out of the flame to lay the bead.
9. The bead is formed by depositing the metal along a fluxed path across the metal.
10. Lay additional beads at 1" intervals until you can lay complete beads that are straight and even.
11. Clean the area and put away the materials.

Instructor's Score or Approval _____

Job Sheet 24-6

Name _____ Date _____

Brazing

Objective

When you have completed this activity you will have demonstrated the ability to braze.

Tools and Materials

Job sheet 24-6

Three pieces of flat steel, 1/8" x 2" x 2"

Emery cloth

Face shield

Leather gloves

Leather apron

Welding goggles with shade 5 lenses

Brazing rod

Flux

Oxyfuel torch

Procedure

1. Obtain permission.
2. Put on the protective clothing.
3. Clean 1" of the edge of each of two pieces.
4. Lay the two steel pieces on the firebricks so clean 1/2" strips overlap, using the third piece to support the second.
5. Clamp the assembly or hold it in place with a brick.
6. Light and adjust the torch.
7. Apply heat until both pieces are dull red.
8. Add flux by way of the rod.
9. Continue to heat the metal to a dull red and move the rod in and out to deposit metal.
10. Play the heat on the upper piece and back from the edge to draw the filler into the joint.
11. Allow the assembly to cool.
12. Turn the assembly over and check for completeness.
13. Clean the area and put away the materials.

Instructor's Score or Approval _____

> **Job Sheet 24-7**

Name _____ Date _____

Braze Welding Butt Welds

Objective
When you have completed this activity you will have demonstrated the ability to braze weld a butt joint.

Tools and Materials
Job sheet 24-7

Two pieces of 16 gauge steel x 2" x 2"

Oxyfuel torch

Filler rod

Flux

Fine emery cloth

Shade 5 welding goggles

Face shield

Leather gloves

Leather apron

Firebricks

Procedure
1. Obtain permission.
2. Put on the protective clothing.
3. Clean the metal with the emery cloth.
4. Place them on the flat firebrick with a 1/32" gap between them.
5. Weight the two pieces with firebricks to hold them in position.
6. Light and adjust the torch.
7. Heat a spot at one end of the joint, flux the area, and deposit a spot of filler material.
8. Repeat the tack at the other end of the joint.
9. Lay a bead over the entire length of the joint.
10. Clean the area and put away the materials.

Instructor's Score or Approval _____

Job Sheet 24-8

Name _____ Date _____

Braze Welding Fillet Welds

Objective

When you have completed this activity you will have demonstrated the ability to do fillet braze welding.

Tools and Materials

- Job sheet 24-8
- Oxyfuel torch
- Firebricks
- Emery cloth
- Two pieces of 16 gauge steel x 1 1/2" x 6"
- Welding goggles with shade 5 lens
- Face shield
- Leather gloves
- Leather apron
- Filler rod
- Flux

Procedure

1. Obtain permission.
2. Put on the protective clothing.
3. Clean the metal with the emery cloth.
4. Place one piece flat on the firebrick.
5. Place the second piece on its edge on the first and prop them in this position.
6. Light and adjust the torch.
7. Tack weld the ends and middle.
8. Lay a bead all along the joint.
9. Clean the area and put away the materials.

Instructor's Score or Approval _____

> **Job Sheet 24-9**

Name _____ Date _____

Pushing the Puddle

Objective

When you have completed this activity you will have demonstrated the ability to push the puddle.

Tools and Materials

  Job sheet 24-9

  Oxyfuel torch

  Firebricks

  Shade 5 welding goggles

  Face shield

  Leather gloves

  Leather apron

  16 gauge x 2" x 6" steel

  Emery cloth

Procedure

1. Obtain permission.
2. Put on the protective clothing.
3. Clean the metal with the emery cloth.
4. Weight the metal on a firebrick.
5. Light and adjust the torch.
6. Start at one end and point the tip at a 45 degree angle in the direction of travel and with a gap of 1/8" between tip and metal.
7. When the metal starts to melt, move the torch in a small circular pattern to form the bead.*
8. Lift the torch slowly at the end of the bead.
9. Repeat beads until the plate is full.
10. Clean the area and put away the materials.

    *Examine text Figure 24-15 to compare your beads.

Instructor's Score or Approval _____

Job Sheet 24-10

Name _____ Date _____

Making a Corner Weld Without Filler Rod

Objective

When you have completed this activity you will have demonstrated the ability to make a corner weld without filler rod.

Tools and Materials

    Job sheet 24-10

    Oxyfuel torch

    Shade 5 welding goggles

    Face shield

    Leather gloves

    Leather apron

    Emery cloth

    Two pieces of steel 16 gauge x 1 1/2" x 6"

    Firebricks

Procedure

1. Obtain permission.
2. Put on the protective clothing.
3. Clean the edges of the metal with the emery cloth.
4. Put the two together so their edges make a 60 degree tent.
5. Light and adjust the torch.
6. Tack weld the ends and two places between.
7. Make a bead all along the joint.
8. Clean the area and put away the materials.

Instructor's Score or Approval _____

SECTION 8 ARC WELDING

Unit 25 Selecting and Using Arc Welding Equipment

Arc welding equipment is widely used in agricultural construction and maintenance. Knowledge of arc welding and skill in its use will be valuable to you.

Class Activity 25-1

Understanding Arc Welding

Class Activity 25-2

Identifying Arc Welding Advantages

Class Activity 25-3

Identifying Arc Welding Safety Items

Class Activity 25-4

Identifying Electrode Code Numbers

Job Sheet 25-1

Name _____ Date _____

Understanding Arc Welding

Objective

Upon completing this activity you will have demonstrated an understanding of arc welding.

Tools and Materials

Job sheet 25-1

Pen or pencil

Procedure

Complete the following worksheet.

1. What are the benefits of slag formation in welding?
   _____
   _____

2. The arc temperature can reach _____ degrees F.

3. What are reasons for this temperature to vary?
   _____
   _____

4. What does duty cycle mean when applied to arc welders?
   _____
   _____

5. What are the differences between direct and alternating current welders?
   _____
   _____
   _____
   _____

6. What are the common amperages of popular agricultural welders? _____

7. Portable DC welders are commonly driven by _____
   _____

8. Portable welders may also be used as _____.

9. What happens if too small welding cables are used? _____
   _____

10. Describe a desirable electrode holder. _____
    _____

11. Describe a ground clamp. _____
    _____
    _____

Instructor's Score or Approval _____

Job Sheet 25-2

Name _____ Date _____

Identifying Arc Welding Advantages

Objective

When you have completed this activity you will have demonstrated the ability to identify arc welding advantages.

Tools and Materials

Job sheet 25-2

Pen or pencil

Procedure

Match the items in column I with those in column II.

Column I

___ 1. Inexpensive source of heat for welding

___ 2. Can be used for

___ 3. Can be learned quickly

___ 4. Engine driven

___ 5. Arc welding is

___ 6. Farm arc welders are

Column II

a. portable welders

b. fast and reliable

c. electricity

d. brazing and heating as well as welding

e. arc welding

f. relatively inexpensive

Instructor's Score or Approval _____

Job Sheet 25-3

Name _____ Date _____

Identifying Arc Welding Safety Items

Objective

When you have completed this activity you will have demonstrated the ability to identify arc welding safety items.

Tools and Materials

Job sheet 25-3

Pen or pencil

Procedure

Match the items in column I with those in column II.

Column I

___ 1. Convert power: high voltage low amperage to low voltage high amperage

___ 2. Driven by electric motors or gasoline engines

___ 3. Large cables

___ 4. Welding curtains

___ 5. Spring-loaded device

___ 6. Welding table

Column II

a. DC welders

b. AC welders

c. electrode holders

d. high amperage

e. protection for others

f. comfortable workstation

Instructor's Score or Approval _____

Lab Manual 187

Job Sheet 25-4

Name _____ Date _____

Identifying Electrode Code Numbers

Objective

When you have completed this activity you will have demonstrated the ability to identify the electrode code numbers.

Tools and Materials

Job sheet 25-4

pen or pencil

Procedure

Complete the following worksheet. Remember: The answer pertains only to the number or letter, not to the blanks.

Number		*Meaning*
___ 1. E _ _ _ _		a. thousand pounds of tensile strength
___ 2. _ _ _ _ 1		b. usable in all directions
___ 3. _ _ _ 1 _		c. usable only in the flat or horizontal positions
___ 4. _ _ _ _ 2		d. usable for vertical down only
___ 5. _ _ _ _ 3		e. DC reverse polarity only
___ 6. _ _ _ _ 4		f. AC and DC reverse polarity
___ 7. _ _ _ _ 5		g. AC and DC straight polarity
___ 8. _ _ _ _ 6		h. AC and DC
___ 9. _ _ _ _ 8		i. DC reverse polarity
___ 10. _ _ _ 2 _		j. electrode
___ 11. _ _ _ 4 _		
___ 12. _ 7 0 _ _		
___ 13. _ _ _ _ 0		

Note: Meanings can apply to more than one number.

Instructor's Score or Approval _____

Unit 26 Arc Welding Mild Steel and MIG/TIG Welding

Practice in arc welding can quickly develop skill. In this unit you will practice those skills. Since the use of the arc welder presents a fire danger from the intense heat, you should practice all safety precautions.

Class Activity 26-1

Understanding Arc Welding of Mild Steel

Shop Activity 26-2

Striking the Arc

Shop Activity 26-3

Making a Practice Bead

Shop Activity 26-4

Making a Butt Weld

Shop Activity 26-5

Checking the Butt Weld

Shop Activity 26-6

Making the Fillet Weld

Shop Activity 26-7

Making a Horizontal Weld

Shop Activity 26-8

Making a Vertical Weld

Shop Activity 26-9

Making an Overhead Weld

Shop Activity 26-10

Piercing with the Arc Welder

Shop Activity 26-11

Cutting with the Arc Welder

Shop Activity 26-12

Identifying the Components of a Mig Welder

Class Activity 26-13

Identifying MIG Welding Defects

Job Sheet 26-1

Name _____ Date _____

Understanding Arc Welding of Mild Steel

Objective

Upon completing this activity you will have demonstrated an understanding of arc welding mild steel.

Tools and Materials

Job sheet 26-1

Pen or pencil

Procedure

Complete the following worksheet.

1. Why should the welding area have metal benches?
   _____
   _____

2. The fire extinguisher in the area should be type _____.

3. List the personal protection gear that should be in the welding area.
   _____
   _____
   _____

4. Shade number ____ lenses should be used in arc welding.

5. The electrode angle recommended is _____ degrees.

6. Why is sound a good guide to arc length?
   _____
   _____
   _____

7. What are the best indicators of proper speed of travel?
   _____
   _____
   _____

8. What happens when the amperage is set too low?
   _____
   _____
   _____

9. When the amperage is set too high?
   _____
   _____
   _____

Instructor's Score or Approval _____

Job Sheet 26-2

Name _____ Date _____

Striking the Arc

Objective

When you have completed this activity you will have demonstrated the ability to strike an arc.

Tools and Materials

Job sheet 26-2

Arc welder

E6011 electrodes

Chipping hammer

Welding helmet with shade 10 lens

Welding table

Mild steel scrap or 1/8" to 1/4" x 2" x 6"

Leather gloves

Leather apron

Face shield unless the welding helmet has a flip-up lens

Procedure

1. Obtain permission.
2. Put on the leather gloves and apron.
3. Set the arc welder to 80 to 100 amps.
4. Ground the metal to be used.
5. Put an electrode in the electrode holder.
6. Turn on the welder.
7. Lower the helmet into position.
8. Lower the electrode from 1/8" above the metal to touch the metal and then lift it quickly like striking a match.
9. If you are too slow in lifting and the electrode sticks, twist it free or release the holder.
10. If the coating on the electrode is ruined, put it aside and use another (this damaged electrode can be used by a more experienced welder).
11. If you lift too high and the arc is broken, try again and stay closer to the metal.
12. Feed the electrode down slowly as it burns away, travel 1" and break the arc.
13. Practice steps 8 to 12 until you have 16 1" beads on your metal.
14. Compare your beads to Figure 26-10.
15. Clean the area and put away the materials.

Instructor's Score or Approval _____

Job Sheet 26-3

Name _____ Date _____

Making a Practice Bead

Objective

When you have completed this activity you will have demonstrated the ability to run a bead.

Tools and Materials

Job sheet 26-3

Arc Welder

E6011 electrodes

Welding helmet with shade 10 lens

Leather gloves

Leather apron

Welding table

Chipping hammer

Mild steel 1/4" x 4" x 6"

Welding clamp

Face shield unless the helmet is equipped with a flip-up lens

Procedure

1. Obtain permission.
2. Put on the leather gloves and apron.
3. Clamp the metal to the welding table.
4. Set the amperage to 110 amps.
5. Put the electrode in the holder.
6. Turn on the welder and lower the helmet.
7. Lean the electrode about 80 degrees and strike the arc.
8. Move quickly to the starting spot at the edge of the plate.
9. Feed the electrode down slowly, watching the puddle and moving to the right.
10. At the end of the bead, lift the electrode slowly until the center of the puddle fills and the arc goes out.
11. Clean the bead with the chipping hammer using the flip-up clear lens or a face shield.
12. Compare your bead with those in text Figure 26-11.
13. Repeat until you can make excellent beads.
14. Lower the amperage if the pad gets too hot.
15. Clean the area and put away the materials.

Instructor's Score or Approval _____

> **Job Sheet 26-4**

Name _____ Date _____

Making a Butt Weld

Objective

When you have completed this activity you will have demonstrated the ability to make a butt weld.

Tools and Materials

  Job sheet 26-4
  Arc welder
  Welding table
  Chipping hammer
  Welding helmet with shade 10 lens
  Grinder
  Leather gloves
  Leather apron
  Two pieces of mild steel 1/4" x 2" x 6"
  E6011 electrodes
  Face shield unless the helmet has a flip-up lens

Procedure

1. Obtain permission.
2. Put on the protective clothing.
3. Grind one edge of each piece to 1/8".
4. Place the pieces 1/8" apart.
5. Set the amperage to 110.
6. Turn on the welder and lower the helmet.
7. Strike the arc and move quickly to tack each end.
8. Flatten the pieces if needed.
9. Create an even 1/4" wide bead.
10. Chip the slag with the chipping hammer.
11. Examine the weld; the answers to these questions should be yes.
    a. Do the sides blend in evenly with the base metal?
    b. Are the semicircles of the bead evenly spaced?
    c. Does the weld go all the way through the metal?
    d. Does the weld start at the starting edge?
    e. Does the weld fill the groove at the start?
    f. Does the weld go all the way to the finishing edge?
    g. Does the weld fill the groove at the finishing edge?
12. Clean the area and put away the materials.

Instructor's Score or Approval _____

Job Sheet 26-5

Name _____ Date _____

Checking the Butt Weld

Objective

When you have completed this activity you will have demonstrated the ability to check a butt weld.

Tools and Materials

Job sheet 26-5

Leather gloves

Leather apron

Face shield

Hacksaw

Machinist's vise

Two pound hammer

Procedure

1. Saw 1" off the end of the butt weld completed as Job Sheet 26-4.
2. Examine the cross section for voids or slag.
3. Repeat steps 1 and 2.
4. Clamp the remainder in a heavy vise.
5. With a heavy hammer, attempt to break the weld.
6. If the weld breaks, determine the problem by comparing your weld with the text and conferring with your instructor.
7. Repeat the practice until you produce good welds.
8. Clean the area and put away the materials.

Instructor's Score or Approval _____

Job Sheet 26-6

Name _____ Date _____

Making the Fillet Weld

Objective

When you have completed this activity you will have demonstrated the ability to make a fillet weld.

Tools and Materials

Job sheet 26-6

Arc welder

Welding helmet with shade 10 lens

Filler rod

Flux

Chipping hammer

Leather gloves

Leather apron

E6011 electrodes

Face shield

Two pieces of steel 1/8" x 1" x 6"

Welding clamp

Face shield

Procedure

1. Obtain permission.
2. Clamp one piece vertically on a base piece.
3. Set the welder to 110 amps.
4. Tack weld both ends.
5. Square the piece to 90 degrees.
6. Hold the electrode at 45 degrees and lean it in the direction of travel.
7. Run the bead all the way along the joint.
8. Keep the penetration equal by controlling the tip of the electrode.
9. Run a bead on the other side, alternating passes to control distortion.
10. Chip and examine the weld.
11. Clean the area and put away the materials.

Instructor's Score or Approval _____

Job Sheet 26-7

Name _____ Date _____

Making a Horizontal Weld

Objective

When you have completed this activity you will have demonstrated the ability to make a horizontal weld.

Tools and Materials

    Job sheet 26-7

    Arc welder

    E6011 electrodes

    1/8" or 1/4" x 4" x 6" steel

    Chipping hammer

    Welding helmet with shade 10 lens

    Leather gloves

    Leather arm and shoulder protection

    Leather apron

    Welding table

    Face shield

Procedure

1. Obtain permission.
2. Put on the protective clothing.
3. Set the welder to 110 amps.
4. Weld two small pieces to the pad so it can be set on edge and clamped to the table.
5. Sit in front of the table.
6. Start at the left end with the electrode at 70 degrees to the plate and tilted so it tips upward at about 80 degrees.
7. Run the bead at a speed that creates a proper bead without sagging or undercutting.
8. Adjust the amperage and run more beads to develop the skill.
9. Clean the area and put away the materials.

Instructor's Score or Approval _____

Job Sheet 26-8

Name _____ Date _____

Making a Vertical Weld

Objective

When you have completed this activity you will have demonstrated the ability to make a vertical weld.

Tools and Materials

    Job sheet 26-8

    Arc welder

    Welding table

    Clamps

    E6011 electrodes

    Welding helmet with shade 10 lens

    Chipping hammer

    Leather gloves

    Leather apron

    Leather shoulder and arm protection

    Face shield

    1/8" or 1/4" x 4" x 6" practice pad

Procedure

1. Obtain permission.
2. Put on the protective clothing.
3. Set up the practice pad as in Job Sheet 26-7.
4. Start at the bottom and work up.
5. Use a wide weave pattern to form the bead.
6. If the weld is too hot, stop and clean the weld then adjust the amperage until control is achieved.
7. Practice until you can make a sound bead.
8. For a vertical-up fillet weld, use a T-pattern weave from one side to the other.
9. Clean the area and put away the materials.

Instructor's Score or Approval _____

Job Sheet 26-9

Name _____ Date _____

Making an Overhead Weld

Objective

When you have completed this activity you will have demonstrated the ability to make an overhead weld.

Tools and Materials

    Job sheet 26-9

    Arc welder

    Helmet with shade 10 lens

    Leather arm and shoulder protection

    Leather gloves

    Leather apron

    E6011 electrodes

    Practice pad

    Clamps

    Welding table

    Face shield

Procedure

1. Obtain permission.
2. Clamp the pad in an overhead position.
3. Put on the protective clothing.
4. Set the amperage to 100.
5. Hold the electrode nearly straight up.
6. Run the bead adjusting amperage as necessary.
7. Practice until you develop the skill.
8. Clean the area and put away the materials.

Instructor's Score or Approval _____

Job Sheet 26-10

Name _____ Date _____

Piercing with the Arc Welder

Objective

When you have completed this activity you will have demonstrated the ability to pierce with the arc welder.

Tools and Materials

- Job sheet 26-10
- Arc welder
- Helmet with shade 10 lens
- E6011 electrodes
- Welding table
- Clamps
- Metal to pierce
- Bucket of sand
- Leather gloves
- Leather apron
- Leather foot and leg protection

Procedure

1. Obtain permission.
2. Clamp the metal so it extends 4" over the edge of the table.
3. Place the bucket of sand to collect the hot metal.
4. Put on the protective clothing.
5. Set the welder to 150 amperes.
6. Strike and hold a long arc until the metal melts.
7. Quickly poke the electrode through the molten metal.
8. If it sticks, whip it loose or start again.
9. Wait longer to poke the electrode into the puddle.
10. Move the electrode around the hole to enlarge it as needed.
11. If the hole must be round, put it over the hole in an anvil and drive a punch through the hot metal.
12. Smooth the underside with the grinder.
13. Clean the area and put away the materials.

Instructor's Score or Approval _____

Job Sheet 26-11

Name _____ Date _____

Cutting with the Arc Welder

Objective

When you have completed this activity you will have demonstrated the ability to cut with the arc welder.

Tools and Materials

    Job sheet 26-11

    Arc welder

    Helmet with shade 10 lens

    Welding table

    Bucket of sand

    Clamps

    Metal to be cut

    E6011 electrodes

    Leather gloves

    Leather apron

    Leather foot and leg protection

Procedure

1. Obtain permission.
2. Clamp the metal so it extends over the table.
3. Position the sand to catch the hot metal.
4. Put on the protective clothing.
5. Set the welder to 150 amperes.
6. Hold a long arc at the edge of the metal until it is molten.
7. Use quick up and down motions to chop the metal away and create a kerf across the metal.
8. Smooth with the grinder as needed.
9. Clean the area and put away the materials.

Instructor's Score or Approval _____

200 Lab Manual

Job Sheet 26-12

Name _____ Date _____

Identifying the Components of a MIG Welder

Objective

When you have completed this activity you will have demonstrated the ability to identify the components of the MIG welder.

Tools and Materials

　　Job sheet 26-12

　　Pen or pencil

Procedure

　　Complete the following worksheet by identifying the components of the MIG welder.

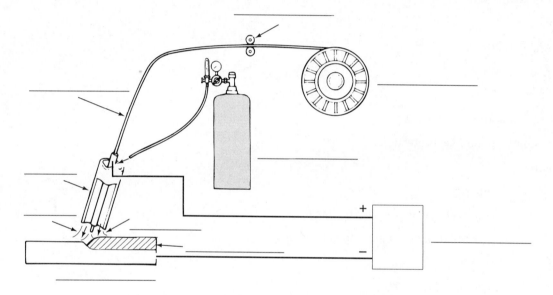

Figure 26-1. Components of the MIG Welder

Instructor's Score or Approval _____

Lab Manual 201

Job Sheet 26-13

Name _____ Date _____

Identifying MIG Welding Defects

Objective

When you have completed this activity you will have demonstrated the ability to identify MIG welding defects.

Tools and Materials

Job sheet 26-13

Pen or Pencil

Procedure

Complete the following worksheet.

MIG WELDING DEFECTS	
Defect Condition	**Cause**
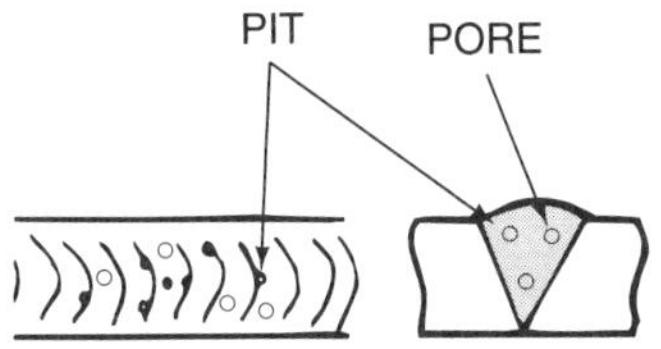 (PIT / PORE diagram)	
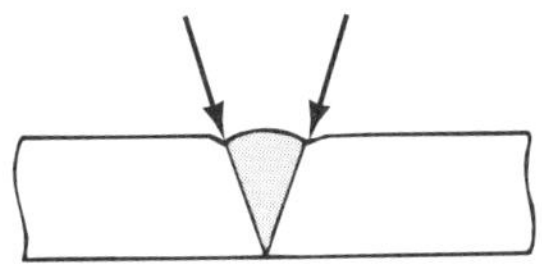 (diagram)	
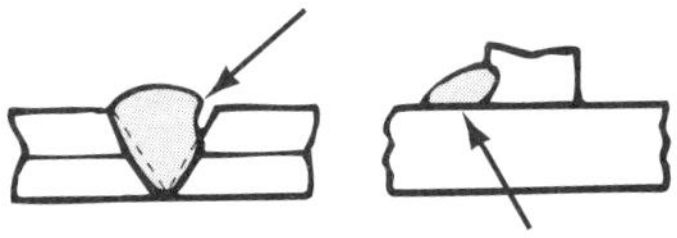 (diagram)	
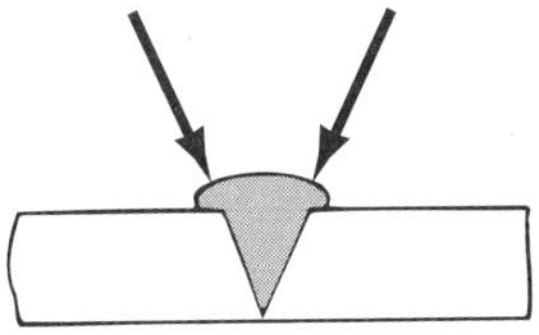 (diagram)	

Figure 26-2. MIG Welding Defects

MIG WELDING DEFECTS	
Defect Condition	**Cause**
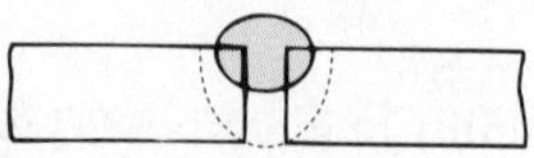	
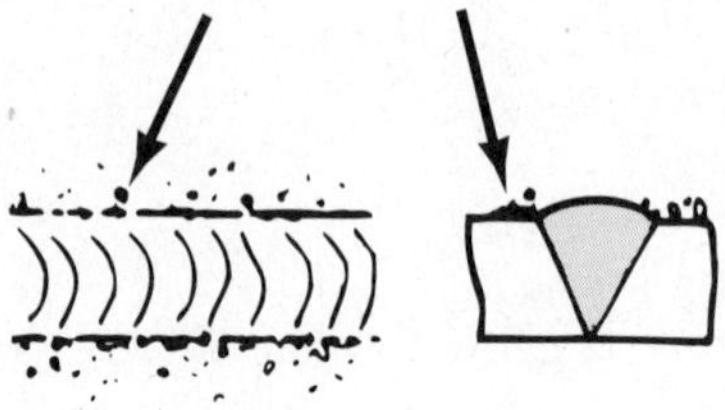	
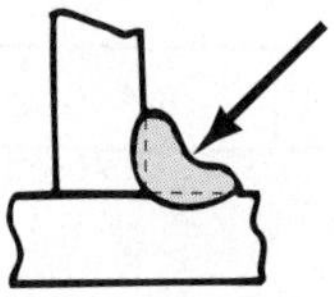	
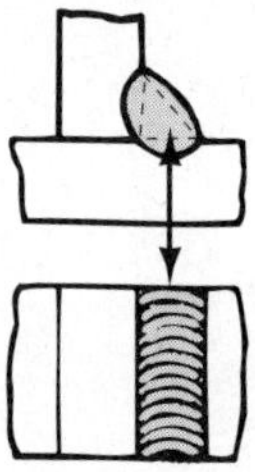	
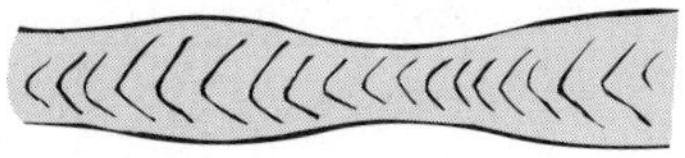	
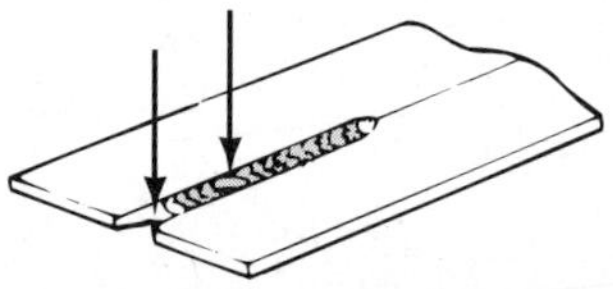	

Figure 26-2. MIG Welding Defects (continued)

Instructor's Score or Approval _____

SECTION 9 PAINTING

Unit 27 Preparing Wood and Metal for Painting

Most agricultural machinery and buildings are exposed to the weather some or all of the time. If metal and wood are exposed to the weather they need to be protected. The most common weather protection is a good coat of paint. In this unit you will learn how to prepare wood and metal for painting.

Class Activity 27-1

Understand Wood and Metal Preparation

Shop Activity 27-2

Preparing New Wood

Shop Activity 27-3

Preparing Old Wood

Shop Activity 27-4

Steam or Pressure Cleaning Machinery

Shop Activity 27-5

Preparing New Metal for Painting

Shop Activity 27-6

Preparing Previously Painted Metal

Shop Activity 27-7

Masking for Spray Painting

Class Activity 27-8

Estimating Paint Jobs

Job Sheet 27-1

Name _____ Date _____

Understand Wood and Metal Preparation

Objective

Upon completing this activity you will have demonstrated an understanding of wood and metal preparation for painting.

Tools and Materials

Job sheet 27-1

Pen or pencil

Procedure

Complete the following worksheet.

1. What are the primary materials used in agricultural buildings and equipment?
   _____
   _____

2. What is the most common protection from moisture damage? _____
   _____

3. Define paint. _____
   _____

4. Define preservatives. _____
   _____

5. What are some common preservatives? _____
   _____

6. Why should creosote and pentachlorophenol be used with caution? _____
   _____
   _____

7. What does the term new wood refer to? _____
   _____

8. What is the purpose of a sealer? _____
   _____
   _____

9. Define primer. _____
   _____

10. Define caulk. _____
    _____

11. How are caulk and primer used? _____
    _____

Instructor's Score or Approval _____

Job Sheet 27-2

Name _____ Date _____

Preparing New Wood

Objective

When you have completed this activity you will have demonstrated the ability to prepare new wood for painting.

Tools and Materials

   Job sheet 27-2

   New wood to be painted

   Stiff bristle brush

   Mineral spirits

   Rags

   Sandpaper

   Household ammonia

Procedure

1. Brush off any dirt from the wood.
2. Clean any grease or oil with a rag dipped in mineral spirits.
3. Clean any wax with household ammonia.
4. Sand to remove any excessively rough spots or to remove dirt not dislodged by the brush.
5. Allow the wood to dry.
6. Clean the area and put away the materials.

Instructor's Score or Approval _____

Job Sheet 27-3

Name _____ Date _____

Preparing Old Wood

Objective

When you have completed this activity you will have demonstrated the ability to prepare old wood for painting.

Tools and Materials

- Job sheet 27-3
- Old wood to be painted
- Paint scrapers
- Wire brush
- Paint and varnish remover
- Mineral spirits
- Rags
- Sandpaper
- Putty or plastic wood

Procedure

1. Remove loose paint with a scraper or wire brush.
2. Remove grease or oil with a rag soaked in mineral spirits.
3. Use paint and varnish remover if a smooth surface is needed.
4. Sand spots needing special attention.
5. Fill all holes with putty or plastic wood.
6. Sand the putty or plastic wood when dry.
7. Clean the area and put away the materials.

Instructor's Score or Approval _____

Job Sheet 27-4

Name _____ Date _____

Steam or Pressure Cleaning Machinery

Objective

When you have completed this activity you will have demonstrated the ability to steam or pressure clean machinery.

Note: The procedure for pressure cleaning omits the steps relating to steam.

Tools and Materials

    Job sheet 27-4

    Steam or pressure cleaner

    Machine to be cleaned

    Cleaning compound

    Water supply

    Plastic bags

    Masking tape

    Fuel for the cleaner

    Safety grounded outlet

    Insulated rubber gloves

Procedure

1. Obtain permission.
2. Cover all electrical contacts on the machine with the plastic bags and tape.
3. Place the machine at least 15 feet from any building.
4. Plug the cleaner motor into a three wire grounded outlet.
5. Attach the water hose to the inlet of the cleaner and the water supply.
6. Fill the fuel tank on the cleaner.
7. Mix the cleaning compound in the solution tank.
8. Position the cleaner so you can monitor all gauges.
9. Lay out the hose.
10. Hold the nozzle away from anyone and toward the concrete or free air.
11. Open the water faucet.
12. Turn on the cleaner motor.
13. Turn on the fuel valve and start the cleaner.
14. Adjust the fuel valve to create the recommended pressure for the cleaner.
15. When steam comes from the nozzle, adjust the cleaning solution to obtain the desired cleaning action.
16. Start at the top of the machine and hold the nozzle about 8″ to 12″ from the machine.
17. Use 2″ to 4″ for tougher places.
18. Steam the whole machine lightly to remove any dirt.
19. Turn off the solution valve.

20. Turn off the fuel valve.
21. Hold the nozzle until cold water comes out.
22. Turn off the water.
23. Unplug the cleaner at the outlet.
24. Disconnect the hose from the machine and hose down the area.
25. Drain the machine and hoses using compressed air if it is to be exposed to freezing temperatures.
26. Properly store the cleaner.
27. Remove the plastic from the machine.
28. Start the machine and dry it.

Instructor's Score or Approval _____

Job Sheet 27-5

Name _____ Date _____

Preparing New Metal for Painting

Objective

When you have completed this activity you will have demonstrated the ability to prepare new metal for painting.

Tools and Materials

    Job sheet 27-5

    Metal to be painted

    Mineral spirits

    Rags

    Commercial solvent

    Emery cloth

    Chipping hammer

    Wire brush

Procedure

1. Chip and brush any welds.
2. Sand to remove any dirt or rust.
3. Clean with mineral spirits to remove grease or oil.
4. Clean with commercial solvent.
5. Clean the area and put away the materials.

Instructor's Score or Approval _____

Job Sheet 27-6

Name _____ Date _____

Preparing Previously Painted Metal

Objective

When you have completed this activity you will have demonstrated the ability to prepare painted metal for painting.

Tools and Materials

    Job sheet 27-6

    Metal to be repainted

    Mineral spirits

    Power sander

    Sandpaper 200 to 600 grit

    Water and pan

    Wire brush

Procedure

1. Obtain permission.
2. Steam or pressure clean if needed.
3. If not steamed, clean with scrapers and solvents.
4. Power sand where required.
5. Feather minor chips by water sanding with very fine grit (200) until no edges can be felt.
6. Sand with 400–600 paper until no sanding marks are visible.
7. Clean thoroughly and dry to prevent rust.
8. Clean the area and put away the materials.

Instructor's Score or Approval _____

Job Sheet 27-7

Name _____ Date _____

Masking for Spray Painting

Objective

When you have completed this activity you will have demonstrated the ability to mask objects for spray painting.

Tools and Materials

    Job sheet 27-7

    Object to be masked

    Masking tape and newspapers

    Grease

    Utility knife

Procedure

1. Use grease on small fittings such as grease zerks.
2. Use newspaper and masking tape to cover glass and gauges.
3. Mask any areas not to be painted by cutting paper 1/4″ less than the size of the area and tape to the edges.
4. Use the utility knife to trim overhanging tape or paper.
5. Mask and tape any areas to be painted a second color.
6. Clean the area and put away the materials.

Instructor's Score or Approval _____

Job Sheet 27-8

Name _____ Date _____

Estimating Paint Jobs

Objective

When you have completed this activity you will have demonstrated the ability to estimate the materials needed for a paint job.

Tools and Materials

Job sheet 27-8

Pen or pencil

Paper

Procedure

A. For buildings:
1. Determine the square footage to be painted.
2. Divide the square footage by the coverage of the paint in square feet.
3. Round up the amount to the next quart.
4. If it results in three quarts, round to a gallon.
5. Allow for rougher spots or needed second coats.
6. If the paint is to be oil based, provide for thinner or brush cleaner.
7. Complete the list of paint materials to be ordered.

B. For machinery:

   Remember: A tractor takes a gallon of the main color and a quart of the trim paint.
   1. Compare the size and area of the machine to a small tractor.
   2. Use the same rounding of materials as with the previous exercise.
   3. Order thinner in proportion to the amount of paint.
   4. Order less expensive materials for cleaning the spray equipment.
   5. Complete your list of materials.

Instructor's Score or Approval _____

Unit 28 Selecting and Applying Painting Materials

In the last unit you learned how to prepare surfaces for paint. In this unit we are concerned with selecting and applying paint to beautify or protect wood or metal surfaces.

Class Activity 28-1

Understanding Painting Materials

Class Activity 28-2

Identifying Painting Materials

Shop Activity 28-3

Using Brushes

Shop Activity 28-4

Using Rollers

Shop Activity 28-5

Using Aerosols

Shop Activity 28-6

Using Spray Guns

Shop Activity 28-7

Cleaning Spray Equipment

Job Sheet 28-1

Name _____ Date _____

Understanding Painting Materials

Objective

Upon completing this activity you will have demonstrated an understanding of painting materials.

Tools and Materials

Job sheet 28-1   Pen or pencil

Procedure

1. What are the two main reasons for painting?
   _____
   _____
   _____

2. What do we mean by formulation?
   _____
   _____
   _____

3. What is pigment?
   _____
   _____
   _____

4. What differentiates high from low quality pigments?
   _____
   _____
   _____

5. What are primers or undercoats? _____
   _____
   _____
   _____

6. What do we mean by compatibility of paints? _____
   _____

7. What is rust resistance in paints?
   _____
   _____
   _____

8. Name the parts of the spray gun pictured below.

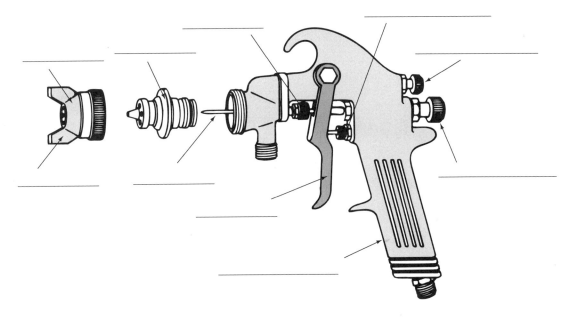

Figure 28-1. Parts of a Spray Gun

Instructor's Score or Approval _____

Job Sheet 28-2

Name _____ Date _____

Identifying Painting Materials

Objective

When you have completed this activity you will have demonstrated the ability to identify painting materials.

Tools and Materials

Job sheet 28-2

Pen or pencil

Procedure

Match the items in column I with those in column II.

Column I

___ 1. Alkyd
___ 2. Latex
___ 3. Thinner
___ 4. Epoxy
___ 5. Interior
___ 6. Exterior
___ 7. Pigment
___ 8. Hiding power
___ 9. Gloss
___ 10. Semigloss
___ 11. Flat
___ 12. Primer
___ 13. Vehicle

Column II

a. without shine
b. special adhesive and wear qualities
c. material that carries the color in painting
d. gives color to paint
e. thins paint
f. has a light shine
g. shiny
h. prepares surface for a high quality topcoat
i. suitable for inside use
j. ability to cover previous colors
k. formulated for weather exposure
l. water cleanup
m. oil-based

Instructor's Score or Approval _____

Job Sheet 28-3

Name _____ Date _____

Using Brushes

Objective

When you have completed this activity you will have demonstrated the ability to use paint brushes.

Tools and Materials

Job sheet 28-3

Paint

Brushes

Brush cleaner

Drop cloths

Paint stirrer

Masking tape

Newspaper

Procedure

1. Prepare the surface to be painted.
2. Mask any parts not to be painted.
3. Put drop cloths below the object to be painted.
4. Use masking tape where the paint is to stop.
5. Mix the paint thoroughly.
6. For oil-based paints use a natural bristle brush and artificial bristle brush for latex paints.
7. Select the brush to fit the job.
8. Dip the bristles 1/3 of the way into the paint and touch the side to remove excess paint.
9. Touch the brush several places on the area to be painted to deposit the paint.
10. Using the flat side of the brush smooth the paint, ending strokes in the painted area.
11. Stir the paint frequently.
12. Paint adjacent areas before the paint dries so the paint coat is uniform.
13. When painting is interrupted, wrap the brush in a plastic bag to prevent the paint hardening in the brush.
14. At the end of the day, clean latex brushes with soap and water and wrap them in paper towels to dry.
15. Oil-based paint brushes may be suspended in paint thinner or linseed oil overnight.
16. When resuming painting, wipe excess thinner or linseed oil from the brushes with rags.
17. At the completion of the job, clean the brush with paint thinner or brush cleaner.
18. Clean the area and put away the materials.

Instructor's Score or Approval _____

Job Sheet 28-4

Name _____ Date _____

Using Rollers

Objective

After completing this activity you will have demonstrated the ability to paint with a roller.

Tools and Materials

- Job sheet 28-4
- Paint tray
- Roller
- Aluminum foil or tray liner
- Paint
- Drop cloths
- Object to be painted
- Small brush

Procedure

1. Line the tray with the aluminum foil or a tray liner.
2. Mix the paint.
3. Place the drop cloths.
4. Pour paint into the tray.
5. Paint corners and edges with the small brush first.
6. Roll the larger areas using a short nap roller for smooth surfaces or longer nap roller for rougher surfaces.
7. Finish by rolling toward the painted area.
8. Stir the paint before refilling the pan.
9. If painting is stopped, wrap the roller in an airtight plastic bag.
10. When done with latex paint, clean quality rollers with soap and water.
11. With oil-based paint or low quality rollers, discard the covers.
12. Clean the roller handle.
13. Remove and discard the pan liner and clean any remaining paint.
14. Clean the small brush and wrap it to dry.
15. Clean the area and put away the materials.

Instructor's Score or Approval _____

Job Sheet 28-5

Name _____ Date _____

Using Aerosols

Objective

When you have completed this activity you will have demonstrated the ability to use aerosol paints.

Tools and Materials

Job sheet 28-5

Object to be painted

Aerosol paint

Cardboard

Drop cloth

Masking

Procedure

1. Prepare the object to be painted.
2. Mask as needed.
3. Choose a well-ventilated area for painting.
4. Shake the aerosol can for several minutes.
5. Practice on the cardboard to see the pattern.
6. Spray in rapid uniform strokes starting in the air and moving to the object.
7. Use a mist coat and when tacky build up the thickness needed.
8. When finished, invert the can and spray until no paint emerges.
9. If more coats are required, follow the directions on the can.
10. Clean the area and put away the materials.

Instructor's Score or Approval _____

Job Sheet 28-6

Name _____ Date _____

Using Spray Guns

Objective

When you have completed this activity you will have demonstrated the ability to use a spray gun in painting.

Tools and Materials

    Job sheet 28-6

    Spray equipment

    Paint

    Thinner

    Paint strainer

    Rags

    Object to be painted

    Safety glasses

    Paint respirator

Procedure

1. Prepare and mask the object to be painted.
2. Mix the paint thoroughly.
3. Test the thinner with a small amount of paint to be sure they are compatible.
4. Thin the paint to the recommended viscosity.
5. Strain the paint into the cup.
6. Put on the safety glasses and the paint respirator.
7. Turn on the air and set the regulator to the recommended pressure for the gun.
8. Adjust the gun using a cardboard target.
9. Use uniform strokes parallel to the surface and 6" to 10" from the surface to obtain a uniform covering without dusting or sagging.
10. Paint the object using passes that overlap 50%.
11. Clean the area and put away all the materials but the spray gun, then use the next Job Sheet to clean the spray gun.

Instructor's Score or Approval _____

Job Sheet 28-7

Name _____ Date _____

Cleaning Spray Equipment

Objective

When you have completed this activity you will have demonstrated the ability to clean the spray equipment.

Tools and Materials

    Job sheet 28-7

    Spray equipment used in the previous activity

    Paint thinner

    Rags

    Cardboard

Procedure

1. Pour all unused paint back into the can and reseal it.
2. Wipe most of the paint from the gun.
3. Pour 1" of paint thinner into the gun.
4. Shake it vigorously.
5. Spray the paint thinner onto the cardboard.
6. Wipe all visible surfaces with paint thinner on a cloth.
7. Repeat steps 3 to 6 once or twice until no traces of paint appear.
8. Remove the air and paint nozzle.
9. Use a cloth and thinner to wipe away any traces of paint.
10. Dry all parts and reassemble the gun.
11. Clean the rest of the equipment with solvent and cloth.
12. Discard cleaning rags and cans in an approved metal container.
13. Clean the area and put away the materials.

Instructor's Score or Approval _____

SECTION 10 SMALL GAS ENGINES

Unit 29 Fundamentals of Small Engines

Small engines are a very common source of power in agricultural mechanics. Small engines also are used for many applications by home owners in urban areas. Knowledge of their fundamentals is the subject of this unit.

Class Activity 29-1

Understanding Small Engine Fundamentals

Class Activity 29-2

Identifying Safety Precautions with Small Engines

Job Sheet 29-1

Name _____ Date _____

Understanding Small Engine Fundamentals

Objective

Upon completing this activity you will have demonstrated an understanding of small engine fundamentals.

Tools and Materials

    Job sheet 29-1

    Pen or pencil

Procedure

Complete the following worksheet.

1. What is an external combustion engine?
   _____
   _____
   _____

2. What is an internal combustion engine?
   _____
   _____
   _____

3. Why was the development of the internal combustion engine so important to agriculture?
   _____
   _____

4. Name the four strokes of the four cycle engine and describe what is happening in each.
   First stroke _____
   _____
   _____

Second stroke _____

_____

Third stroke _____

_____

Fourth stroke _____

_____

5. What are the major differences with a two-cycle engine?

_____

_____

6. What are the advantages of the two-cycle engine?

_____

7. What is compression?

_____

8. How is it provided for in the cylinder?

_____

9. What are the functions of the valves?

_____

10. Why must they fit tightly?

_____

11. Name the parts of the fuel system.

_____

12. Describe the venturi.

_____

13. Name the part of the ignition system.

_____

14. What is a governor?
_____
_____

15. Why is a cooling system needed?
_____

16. What are the four functions of oil ini a small engine?
    a. _____
    b. _____
    c. _____
    d. _____

Instructor's Score or Approval _____

Job Sheet 29-2

Name _____ Date _____

Identifying Safety Precautions with Small Engines

Objective

When you have completed this activity you will have demonstrated the ability to identify safety precautions with small engines.

Tools and Materials

Job sheet 29-2

Pen or pencil

Procedure

Match the items in column I with those in column II.

	Column I		Column II
___ 1.	Never start an engine	a.	overload the engine
___ 2.	Avoid smoking	b.	handle gasoline outdoors
___ 3.	Always wear	c.	as a cleaner
___ 4.	Keep hands and feet	d.	around gasoline or oil
___ 5.	Stop the engine and cool	e.	above recommended speeds
___ 6.	Avoid spills	f.	away from moving parts
___ 7.	Whenever possible	g.	before making adjustments
___ 8.	Disconnect spark plug	h.	of gasoline on hot parts
___ 9	Do not operate engines	i.	in an unventilated area
___ 10.	Do not	j.	before refueling
___ 11.	Never use gasoline	k.	safety glasses and leather shoes

Instructor's Score or Approval _____

Unit 30 Small Engine Maintenance and Repair

Small engines provide low horsepower, portable power units for many agricultural and recreational uses. This unit is designed to help you learn to maintain and repair small engine units.

Class Activity 30-1

Understanding Small Engine Maintenance and Repair

Class Activity 30-2

Identifying Safety Precautions

Shop Activity 30-3

Servicing Foam Air Cleaners

Shop Activity 30-4

Servicing Dry Element Air Cleaners

Shop Activity 30-5

Servicing Duel Element Air Cleaners

Shop Activity 30-6

Servicing Oil Bath Air Cleaners

Shop Activity 30-7

Adjusting the Carburetor

Shop Activity 30-8

Changing Crankcase Oil

Shop Activity 30-9

Troubleshooting

Shop Activity 30-10

Servicing Starter Mechanisms

Shop Activity 30-11

Preparing Small Engines for Storage

Shop Activity 30-12

Deciding the Big Question; Repair or Replace?

Job Sheet 30-1

Name _____ Date _____

Understanding Small Engine Maintenance and Repair

Objective

Upon completing this activity you will have demonstrated an understanding of small engine maintenance and repair.

Tools and Materials

Job sheet 30-1

Pen or pencil

Procedure

Complete the following worksheet.

1. List the basic small engine maintenance tools required. _____
   _____
   _____
   _____
   _____

2. T F Proper attention to maintenance will usually provide years of service.
3. T F Air cleaners are important because they prevent air pollution.
4. T F Air cleaners need only be serviced at the 100-hour interval.
5. T F All air cleaners are serviced the same way.
6. T F Any damaged air cleaner element should be replaced.
7. T F Clean fuel is essential to proper operation.
8. T F Gasoline that sits for a period of time may lose some of its value.
9. T F A partially sheared flywheel key can cause starting problems.
10. T F A valve face can be ground to 1/4 of its margin.
11. T F Starting mechanisms require no maintenance.
12. T F Plastigage is used to measure compression.
13. T F Carbon deposits can be caused by blow-by.
14. T F Major parts may cost more to replace than replacing the engine.
15. T F Small engine troubleshooting skills are useful to many people.

Instructor's Score or Approval _____

Job Sheet 30-2

Name _____ Date _____

Identifying Safety Precautions

Objective

When you have completed this activity you will have demonstrated the ability to identify safety precautions

Tools and Materials

Job sheet 30-2

Pen or pencil

Procedure

Match the items in the column I with those in column II.

Column I *Column II*

___ 1. Work only in a. in the work area

___ 2. Approved fire extinguisher b. in approved metal container

___ 3. Drain fuels c. all oil or fuel spills

___ 4. Store fuels d. found in engine manuals

___ 5. Goggles e. should be kept in work area

___ 6. Wipe immediately f. flammable

___ 7. Avoid flame or sparks g. only in approved containers

___ 8. Observe all safety precautions h. a well-ventilated area

___ 9. Place rags with fuel on them i. wear at all times

___ 10. Keep ___ materials out of the work area j. outdoors

Instructor's Score or Approval _____

Job Sheet 30-3

Name _____ Date _____

Servicing Foam Air Cleaners

Objective

When you have completed this activity you will have demonstrated the ability to clean foam air cleaners.

Tools and Materials

Job sheet 30-3

Small engine with a foam air cleaner

Kerosene

Rags

Motor oil

Procedure

1. Remove the wing nut and cover.
2. Replace the foam if it crumbles or is torn.
3. Wash the foam in the kerosene.
4. Squeeze out the excess liquid.
5. Dry the foam by squeezing it with the cloth.
6. Soak the foam with the recommended motor oil.
7. Squeeze out the excess oil.
8. Clean the metal parts and cover with the solvent.
9. Reassemble the air cleaner.
10. Clean the area and put away the materials.

Instructor's Score or Approval _____

Job Sheet 30-4

Name _____ Date _____

Servicing Dry Element Air Cleaners

Objective

When you have completed this activity you will have demonstrated the ability to service a dry element air cleaner.

Tools and Materials

Job sheet 30-4

Small engine with a dry element air cleaner

Kerosene

Rags

Procedure

1. Remove the wing nut and cover.
2. If the element is clogged or torn, replace it.
3. Tap the element on a hard surface to dislodge dirt.
4. If the recommended hours have elapsed, replace the cartridge with a new one.
5. Clean the area and the cover with the solvent.
6. Reassemble the air cleaner.
7. Clean the area and put away the materials.

Instructor's Score or Approval _____

Job Sheet 30-5

Name _____ Date _____

Servicing Dual Element Air Cleaners

Objective

When you have completed this activity you will have demonstrated the ability to service a dual element air cleaner.

Tools and Materials

    Job sheet 30-5

    Small engine with dual element air filter

    Oil

    Kerosene

    Rags

Procedure

1. Remove the wing nut and cover.
2. Remove the foam precleaner and the dry element.
3. Clean the filters as in Activities 30-3 and 30-4.
4. Reassemble the air cleaner.
5. Clean the area and put away the materials.

Instructor's Score or Approval _____

Job Sheet 30-6

Servicing Oil Bath Air Cleaners

Objective

When you have completed this activity you will have demonstrated the ability to service an oil bath air cleaner.

Tools and Materials

    Job sheet 30-6

    Engine with an oil bath air cleaner

    Oil recommended for the engine

    Kerosene

    Rags

Procedure

1. Remove the wing nut and cover.
2. Remove the core.
3. Remove the cleaner body and dump the dirty oil (remember to dispose of it properly).
4. Clean the body with the kerosene.
5. Reinstall the body and fill with oil to the mark.
6. Rinse the wire element with the solvent
7. Shake the excess solvent from the element.
8. Clean and reinstall the cover and wing nut.
9. Clean the area and put away the materials.

Instructor's Score or Approval _____

Job Sheet 30-7

Name _____ Date _____

Adjusting the Carburetor

Objective

When you have completed this activity you will have demonstrated the ability to adjust a carburetor.

Tools and Materials

    Job sheet 30-7

    Small engine with a carburetor

    Screwdriver

    Tachometer

Procedure

1. Start the engine and warm it to operating temperature.
2. Adjust the idle speed adjustment until the engine idles at the recommended speed as shown on the tachometer.
3. Slowly turn the fuel-air mixture valve clockwise until the engine slows down or labors.
4. Turn the needle slowly counterclockwise until the engine runs smoothly and then slows down.
5. Slowly turn the valve in the other direction until it runs more smoothly.
6. Recheck and adjust the idle speed.
7. Put the engine under load and check its performance.
8. If the engine does not accelerate or pull well, turn the screw slightly to enrich the mixture.
9. Clean the area and put away the materials.

Instructor's Score or Approval _____

> **Job Sheet 30-8**

Name _____ Date _____

Changing Crankcase Oil

Objective

When you have completed this activity you will have demonstrated the ability to change the crankcase oil.

Tools and Materials

    Job sheet 30-8

    Small engine needing an oil change

    Recommended oil

    Adjustable wrench

    Oil drain pan

    Funnel

Procedure

1. Warm the engine.
2. Locate the drain pan under the drain plug and remove the plug.
3. Allow the oil to completely drain.
4. Replace the drain plug.
5. Fill the crankcase with the recommended amount of clean oil.
6. Clean any spills with the rags.
7. Clean the area and put away the materials, disposing of the used oil properly.

Instructor's Score or Approval _____

Job Sheet 30-9

Name _____ Date _____

Troubleshooting

Objective

When you have completed this activity you will have demonstrated the ability to troubleshoot a small engine.

Tools and Materials

    Job sheet 30-9

    Small engine that will not start

    Small engine tool kit

    Safety glasses

Procedure

1. Check that the plug wire is properly installed and has no breaks.
2. Test for spark, with no spark, check all kill switches and wiring for grounding.
3. IWith a weak spark, change the plug.
4. With a good spark, check for a partially sheared flywheel key.
5. Check the fuel tank for fuel.
6. Check the plug for the odor of gasoline after trying to start the engine.
7. If no odor, follow back the lines for any obstruction or fuel filter faults.
8. If the plug is wet, check for flooding due to choke setting or a stuck needle valve.
9. If flooded, let set 10 minutes and try again.
10. If the engin runs but does not produce power, a major overhaul may be needed. (Due to the many different engines available, the procedure for overhaul needs to be followed as given in the operator's manual for the given engine.)
11. Clean the area and put away the materials.

Instructor's Score or Approval _____

Job Sheet 30-10

Name _____ Date _____

Servicing Starter Mechanisms

Objective

When you have completed this activity you will have demonstrated the ability to service starting mechanisms.

Tools and Materials

    Job sheet 30-10

    Small engine with a starter problem

    Repair manual for the engine

    Small engine tools

    Face shield

Procedure

1. Obtain the repair manual for your engine.
2. Follow the repair manual procedure for the type of starter mechanism on your engine. Disassemble the starter carefully because of the spring and sharp edges.
3. Reattach a new starter rope, cleaning all parts.
4. Reassemble the starter mechanism.
5. Clean the area and put away the materials.

Instructor's Score or Approval _____

Job Sheet 30-11

Name _____ Date _____

Preparing Small Engines for Storage

Objective

When you have completed this activity you will have demonstrated the ability to prepare a small engine for storage for a period of time.

Tools and Procedures

    Job sheet 30-11

    Small engine to be prepared for storage

    Recommended oil

    Rags

    Solvent

Procedure

1. Drain the fuel tank.
2. Shut off the fuel valve.
3. Run the engine until all the fuel is burned.
4. Drain the crankcase of a 4-cycle engine and refill with the recommended oil.
5. Remove the spark plug (for 2-cycle engines, rotate the crankshaft until the piston covers the ports) and add 2–3 tablespoons of clean engine oil.
6. Reinstall the spark plug.
7. Rotate the crankshaft several times to distribute the oil.
8. Drain, clean, and reinstall the fuel filter bowl.
9. Remove the shrouds and guards so the engine can be cleaned thoroughly.
10. Clean the exterior.
11. Remove any debris from the fins.
12. Check all linkages for free movement.
13. Loosen all belts.
14. Clean and lubricate all chains.
15. Lubricate all fittings.
16. Reinstall all shrouds and guards.
17. Store the engine off the ground.
18. If outdoors, cover the engine to protect it.
19. Clean the area and put away the materials.

Instructor's Score or Approval _____

Job Sheet 30-12

Name _____ Date _____

Deciding The Big Question: Repair or Replace?

Objective

When you have completed this exercise you will have demonstrated the ability to decide whether to overhaul or replace a small engine.

Tools and Materials

    Job sheet 30-12

    Small engine in need of repair

    Repair manual for the engine

    Small engine tools including specialized items

    Rags

    Kerosene

    Torque wrench

    Plastigage

    Micrometers, inside and outside

Procedures

1. Obtain the manual for your engine.
2. Follow the disassembly procedures including using the ridge reamer to prevent breaking the piston rings upon removing the piston.
3. Use the piston ring expander to remove the piston rings.
4. Using the inside micrometer, measure the cylinder bore parallel and at right angles to the direction of the wrist pin at several different levels within the piston's travel range.
5. Compare the readings with the manufacturer's tolerances.
6. Check the cylinder walls for scoring and wear.
7. Check the crankshaft bearing for scoring and out-of-round with the micrometer.
8. Compare the readings to the tolerances.
9. Cut a piece of Plastigage, put it in the bearing cap, and torque the cap to the specified inch or foot pounds.
10. Measure the flattened Plastigage.
11. Compare the clearance to tolerances.
12. Place the solid piston ring in the cylinder and measure the ring gap clearance with a feeler gauge.
13. Compare the reading to the tolerances.
14. Check the cost of replacement parts needed.
15. Compare the cost of needed parts with the expected wear and the cost of a replacement.
16. Share your comparison measurements with your instructor and agree on a plan of action.
17. If the cost seems reasonable in comparison to expected life, order the parts and reassemble the engine.
18. Clean the area and put away the materials.

Instructor's Score or Approval _____

SECTION 11 ELECTRICITY AND ELECTRONICS

Unit 31 Electrical Principles and Wiring Materials

Electricity is a major source of energy for many operations in agriculture, in the home, and in industry. An understanding of the principles and how electricity works for us will be useful to you in any occupation.

Class Activity 31-1

Understanding Electrical Principles

Shop Activity 31-2

Electric Motor Maintenance

Class Activity 31-3

Identifying Electrical Safety Precautions

Class Activity 31-4

Identifying Standard Electrical Symbols

Job Sheet 31-1

Name _____ Date _____

Understanding Electrical Principles

Objective

Upon completing this activity you will have demonstrated an understanding of electrical principles.

Tools and Materials

   Job sheet 31-1

   Pen or pencil

Procedure

   Complete the following worksheet.

1. _____ is the major power source for stationary equipment in homes, on farms, and in businesses.

2. T F Electricity can produce heat, light, magnetism, and chemical changes.

3. What is an insulator? _____
   _____

4. State Ohm's Law.
   _____
   _____
   _____

5. How are volts, amperes, and watts measured?
   _____
   _____

6. Define magnetism.
   _____
   _____

7. What is a permanent magnet?
   _____
   _____

8. What is the relationship between magnets and electricity?
   _____
   _____
   _____

9. What are magnetic poles?
   _____
   _____

10. Which poles attract each other? _____ poles.

11. Which poles repel each other? _____ poles.

12. Magnetic lines of force are called _____.

13. What are electromagnets? _____
    _____

14. Why are they important? _____
    _____

15. What is an armature? _____
    _____

16. What is a generator? _____
    _____

17. What is an alternator? _____
    _____

18. What periodic maintenance does an electric motor need? _____
    _____

19. What is the best source for information about electric motor maintenance? _____
    _____

20. What is a branch circuit? _____
    _____

21. How are branch circuits protected? _____
    _____

22. What is nonmetallic sheathed cable? _____
    _____

23. What is armored cable? _____
    _____

24. What does wire gauge mean? _____
    _____

25. How can you identify the gauge of a wire? _____
    _____

26. What is the relationship between wire size, length of run, and voltage drop?
    _____

27. Tell the recommended use for each of the following wire types.

Type	Use
T	_____
TW	_____
THHN	_____
THW or THWN	_____
XHHW	_____
UF	_____

Instructor's Score or Approval _____

Job Sheet 31-2

Name _____ Date _____

Electric Motor Maintenance

Objective

When you have completed this activity you will have demonstrated the ability to do electric motor maintenance.

Tools and Materials

    Job sheet 31-2

    Electric motor

    Compressed air source

    Oil

    Solvent

    Owner manual for the motor

    Rags

    Face shield

Procedure

1. Using compressed air, blow out dust and debris from the motor.
2. Using the rag and solvent, clean any remaining dirt or grease from the motor case.
3. Oil the motor according to the owner's manual.
4. Clean the area and put away the materials.

Instructor's Score or Approval _____

Job Sheet 31-3

Name _____ Date _____

Identifying Electrical Safety Precautions

Objective

When you have completed this activity you will have demonstrated the ability to identify safety precautions with electricity.

Tools and Materials

Job sheet 31-3

Pen or pencil

Procedure

Match the items in column I with those in column II.

Column I

___ 1. Safety devices
___ 2. Do not touch
___ 3. Do not remove
___ 4. Use Ground Fault Circuit Interrupters
___ 5. Discontinue use of
___ 6. Do not place
___ 7. Install all wiring to the
___ 8. When a fuse blows or circuit breaker opens
___ 9. Use only double insulated or three-wire grounded cord
___ 10. Do not install
___ 11. Do not leave heat
___ 12. Keep all heaters and lamps away from
___ 13. Keep metal cases of electrical appliances
___ 14. Don't use cracked or broken
___ 15. Follow manufacturer's recommendations
___ 16. Keep appliances dry

Column II

a. in the kitchen and bath and where moisture conditions exist
b. worn electrical cords
c. extension cords under carpets
d. *National Electrical Code*®
e. with wet hands
f. must remain intact
g. third prong on appliance cords
h. for installation of any electrical equipment
i. producing appliances unattended
j. determine the cause before restoring power
k. portable tools
l. higher amperage fuses or circuit breakers than the rating of the circuits they protect
m. combustible materials
n. to reduce shock hazard and prevent rust
o. switches or outlets
p. grounded

Instructor's Score or Approval _____

Job Sheet 31-4

Name _____ Date _____

Identifying Standard Electrical Symbols

Objective

Upon completing this activity you will have demonstrated the ability to identify the standard electrical symbols.

Tools and Materials

Job sheet 31-4

Pen or pencil

Procedure

Complete the worksheet on the next page.

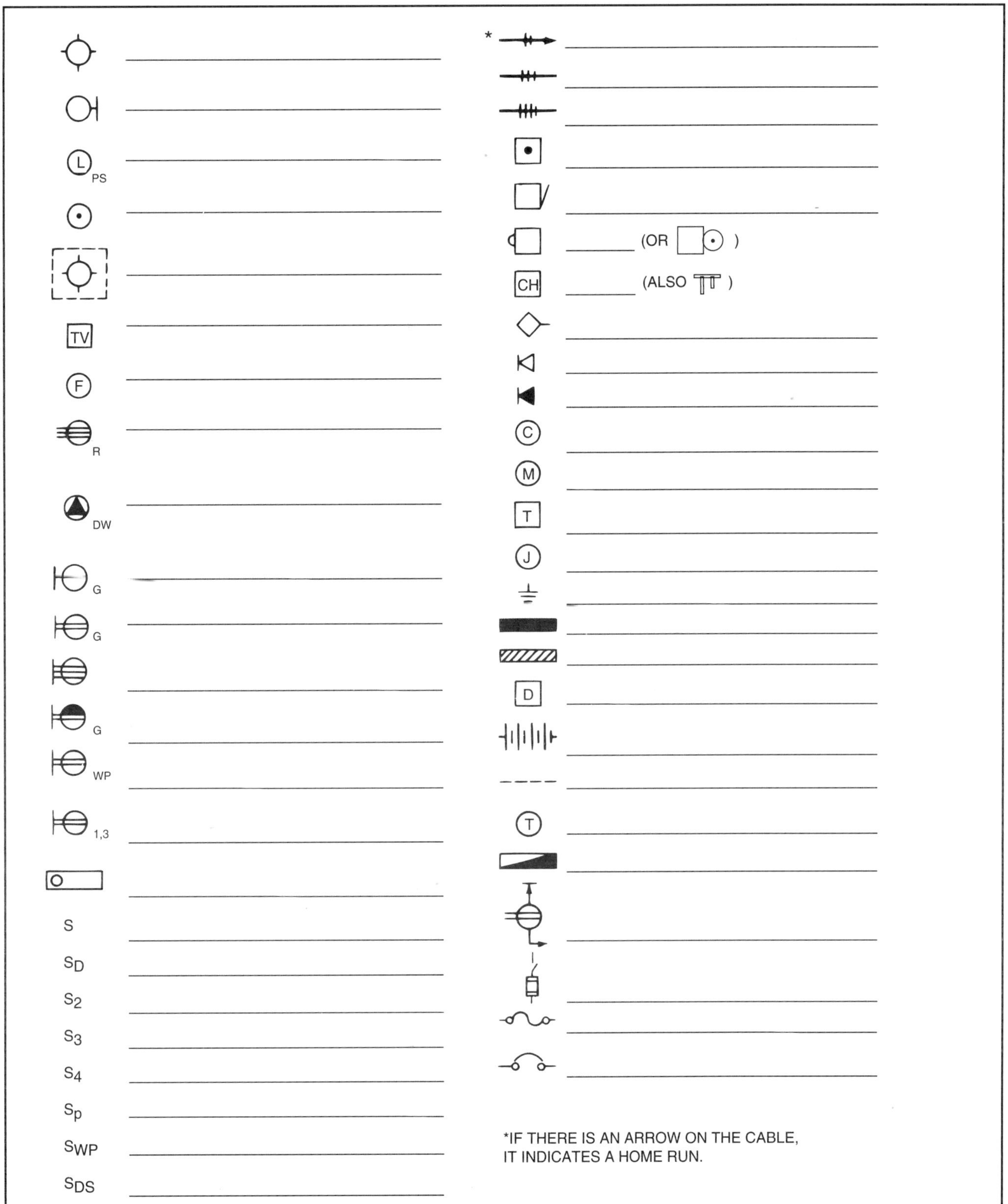
L
PS
TV
F
R
DW
G
G
G
WP
1,3
S
S_D
S_2
S_3
S_4
S_p
S_{WP}
S_{DS}
*
(OR)
CH
(ALSO)
C
M
T
J
D
T
*IF THERE IS AN ARROW ON THE CABLE,
IT INDICATES A HOME RUN.

NOTE: A letter G signifies that the device is of the grounding type. Since all receptacles on new installations are of the grounding type, the notation G is often omitted for simplicity.

Figure 31-1. Electrical Symbols

Instructor's Score or Approval ______________________________

Unit 32 Installing Branch Circuits

All electrical distribution circuits are called branch circuits when they originate from the main entrance panel. The ability to properly wire a branch circuit will be useful in maintaining a home wiring system.

Class Activity 32-1

Understanding Branch Circuits

Shop Activity 32-2

Wiring a Single-Pole Switch and Lamp Holder with the Power Source at the Lamp Holder

Shop Activity 32-3

Wiring a Single-Pole Switch and Lamp Holder with the Power Source at the Switch

Shop Activity 32-4

Wiring a Duplex Outlet

Shop Activity 32-5

Wiring Three-Way Switches and Lamp Holder with the Power Source at One Switch and the Lamp Holder Between the Switches

Shop Activity 32-6

Wiring Three-Way Switches with the Power Source at the Lamp Holder Between the Switches

Shop Activity 32-7

Wiring Three-Way Switches with the Power Source at One Switch and the Lamp Holder Beyond the Switches

Shop Activity 32-8

Wiring a Circuit with One or More Four-Way Switches

Job Sheet 32-1

Name _____ Date _____

Understanding Branch Circuits

Objective

Upon completing this activity you will have demonstrated an understanding of branch circuits.

Tools and Materials

Job sheet 32-1

Pen or pencil

Procedure

Complete the following worksheet.

1. What is the *National Electrical Code*®? Why should all wiring meet this code?

   _____

   _____

   _____

2. What are wiring boxes? Why are they required?

   _____

   _____

3. List the requirements for electrical boxes.

   _____

   _____

   _____

   _____

   _____

4. What is an electrical switch? _____

5. How should a white wire in a switch circuit be marked?

   _____

6. Why should a continuity tester be used to test a circuit before it is turned on?

   _____

   _____

   _____

Instructor's Score or Approval _____

Job Sheet 32-2

Name _____ Date _____

Wiring a Single-Pole Switch and Lamp Holder with the Power Source at the Lamp Holder

Objective

When you have completed this activity you will have demonstrated the ability to wire a single-pole switch and lamp holder with the power source at the lamp holder.

Tools and Materials

Job sheet 32-2
Single-pole switch
Switch cover
Lamp holder
Wall box
Ceiling box
Black electrician's tape

Regular and cross-point screwdrivers
Diagonal cutting pliers
Needle nose pliers
Pocket knife or wire stripper
Wire nuts
Grounding clamps

Procedure

1. Install the wall box and the ceiling box.
2. Install the 14-2 cable with ground from the entrance panel to the ceiling box and from the ceiling box to the wall box.
3. Support and fasten wires according to the *National Electrical Code®*.
4. Slit 6-8″ of the cable cover exposing the wires.
5. At the wall box fasten the bare or green wire to the green screw on the switch. If the box is metal, bundle two 8″ bare wires to the ground wire and attach the second bare wire to a grounding clip on the wall box.
6. Strip 5/8″ of the white and black wires.
7. With the needle nose pliers, make a clockwise loop in the wire's ends.
8. Place the wires under the two screws on the single-pole switch and tighten. Mark the white wire with black tape.
9. Fasten the switch to the wall box with the screw attached.
10. At the ceiling box, fasten the two black wires together with a wire nut and tighten.*
11. Cut an 8″ piece of the bare wire and bundle the three bare wires together and fasten with a wire nut.**
12. Fasten the loose bare wire to the ceiling box with a clip or screw (if the boxes are plastic, no grounding is possible).
13. Wrap a piece of black tape around the white wire coming from the wall box because, in the switch circuit, it becomes a hot wire.
14. Place the formed loop around the brass screw.
15. Place the white wire around the silver or ground screw on the lamp holder.
16. Fasten the lamp holder to the box with the screws provided.
17. Test the circuit with the switch on by touching the white and black leads at the entrance panel with a volt-ohm meter. There should be no reading.
18. Test the circuit with a light bulb installed. There should be a reading.
19. If the boxes are metal, check the boxes and the bare wires for continuity.
20. Install the switch-plate cover.
21. Wire the circuit into the circuit breaker.
22. Clean the area and put away the materials.

   *If the power goes to another outlet or fixture, you will need to have 3 wires in the wire nut.

   **If the power goes on to another outlet or fixture, you will need to have an extra 8″ piece of white wire and bundle three white wires with a wire nut.

Instructor's Score or Approval _____

Job Sheet 32-3

Name _____ Date _____

Wiring a Single-Pole Switch and Lamp Holder with the Power Source at the Switch

Objective

When you have completed this activity you will have demonstrated the ability to wire a single-pole switch with the power source at the switch.

Tools and Materials

Job sheet 32-3	Cross-point and regular screwdrivers
Single-pole switch	Wall box
Lamp holder	Ceiling box
Black electrician's tape	14-2 with ground cable
Volt-ohm meter	Switch-plate cover
Needle nose pliers	Grounding clamps
Diagonal cutting pliers	Wire nuts

Procedure

1. Install the boxes.
2. Install the cables.
3. Strip the cable coverings 8″.
4. Strip 5/8″ of each white and black wire.
5. Make clockwise loops in the black wire ends.
6. Cut an 8″ piece of bare wire and bundle the three bare wires with a wire nut.
7. Ground the loose bare wire to the wall box.
8. Place the white wires in a wire nut and tighten.
9. Place the loops in the black wires around the two terminals on the single-pole switch.
10. Install the switch in the wall box.
11. Clip the bare wire to the ceiling box.
12. Attach the black wire to the brass screw on the lamp holder.
13. Attach the white wire to the silver screw on the lamp holder.
14. Test the circuit with the switch on.
15. Install a lamp bulb and test the circuit again.
16. Test the box's grounding.
17. Connect the circuit to the circuit breaker.
18. Clean the area and put away the materials.

Instructor's Score or Approval _____

Job Sheet 32-4

Name _____ Date _____

Wiring a Duplex Outlet

Objective

When you have completed this activity you will have demonstrated the ability to wire a duplex outlet.

Tools and Materials

Job sheet 32-4

Circuit tester

Duplex outlet

Wall box

Needle nose pliers

Diagonal cutting pliers

12-2 with ground cable

Cross-point and regular screwdrivers

Wire nuts

Grounding clamps

Procedure

1. Install the wall box.
2. Install the cable from the entrance panel to the wall box.
3. Strip 8" of the cable cover.
4. Strip 5/8" of the white and black wires.
5. Cut two 8" pieces of bare wire and bundle them with the bare wire using a wire nut.
6. Ground the metal box with one of the bare copper wires.
7. Make a clockwise loop in the loose bare wire and install it under the green screw.
8. Make the loops and install the black wire under the brass or yellow terminal and the white wire under the silver screw.
9. Install the duplex outlets with the attached screws.
10. Test the circuit for continuity.
11. Wire the circuit to the circuit breaker.*
12. Install the duplex cover.
13. Clean the area and put away the materials.

   *__Note:__ If the circuit is to continue to another outlet, there will be four bare wires in the grounding bundle and a second black wire and a second white wire to attach to the duplex outlet, with the second black wire on the second brass terminal and the second white wire on the second silver terminal.

Instructor's Score or Approval _____

Job Sheet 32-5

Name _____ Date _____

Wiring Three-Way Switches and Lamp Holder with the Power Source at One Switch and the Lamp Holder Between the Switches

Objective

When you have completed this activity you will have demonstrated the ability to wire three-way switches with the lamp holder between the switches.

Tools and Materials

    Job sheet 32-5

    2 three-way switches

    Lamp holder

    2 wall boxes

    Ceiling box

    Wire stripper

    Diagonal cutting pliers

    Needle nose pliers

    Cross-point screwdriver

    Regular screwdriver

    14-2 with ground cable

    14-3 with ground cable

    Wire nuts

    Grounding clamps

Procedure

1. Install the boxes.
2. Install the 14-2 wire from the entrance panel to the first switch.
3. Install the 14-3 wire from the wall boxes to the ceiling box.
4. Strip 8" of the cable covering at each box.
5. Strip 5/8" of each white and black wire.
6. At the power source, cut an 8" piece of bare wire and bundle the three wires with a wire nut.
7. Ground the loose wire to the box.
8. Bundle the two white wires with a wire nut.
9. Attach the black wire from the power source to the single terminal at one end of the switch.
10. Attach the black wire and the red wire from the 3-wire cable to the two terminals on the other end of the switch.
11. Install the switch with the screws provided.
12. At the ceiling box, cut an 8" bare wire and bundle it with the other two wires using a wire nut.
13. Ground the box with the loose bare wire.
14. Attach the two black wires with a wire nut.

15. Attach the two red wires with a wire nut.
16. Attach the white wire from the power source switch to the silver terminal on the lamp holder.
17. Attach the white wire from the second switch (wrap with black tape) to the brass terminal on the lamp holder.
18. Install the lamp holder with the screws provided.
19. At the second switch, ground the bare wire to the box.
20. Attach the red and black wires to the two wires at one end of the switch.
21. Wrap black tape on the white wire and attach it to the single screw on the other end of the switch.
22. Install the switch with the screws provided.
23. Test the circuit for continuity with the switches in both positions and with a light bulb installed.
24. Install the switch covers.
25. Wire the circuit to the circuit breaker.
26. Clean the area and put away the materials.

Instructor's Score or Approval _____

Job Sheet 32-6

Name _____ Date _____

Wiring Three-Way Switches with the Power Source at the Lamp Holder Between the Switches

Objective

When you have completed this activity you will have demonstrated the ability to wire three-way switches with the power source at the lamp holder between the switches.

Tools and Materials

Job sheet 32-6	Black electrician's tape	Diagonal cutting pliers
2 three-way switches	14-3 cable with ground	Cross-point and regular screwdrivers
Lamp holder	14-2 cable with ground	2 switch covers
Two wall boxes	Grounding clamps	Wire nuts
Ceiling box	Needle nose pliers	

Procedure

1. Install the boxes.
2. Install the 14-2 cable to the ceiling box from the entrance panel.
3. Install the 14-3 cable from the switch boxes to the ceiling box.
4. Strip the cables.
5. Strip the wires 5/8".
6. At the ceiling box, cut an 8" piece of bare wire and bundle the three bare wires with a wire nut.
7. Ground the loose bare wire to the box with a grounding clip.
8. Attach the black wire from the power source to the white wire from switch number one, and wrap the white wire with the black tape.
9. Connect the black wires from the switches with a wire nut.
10. Connect the red wires from the switches with a wire nut.
11. Attach the white wire from switch number two to the brass screw on the lamp holder. Mark this wire with black tape.
12. Attach the white wire from the power source to the silver screw on the lamp holder.
13. Install the lamp holder with the screws supplied.
14. At switch number two, ground the bare wire to the box with the grounding clip.*
15. Wrap the white wire with black tape and attach it to the single screw on one end of the switch.
16. Connect the black and red wires to the other two screws.
17. Install the switch with the screws provided.
18. At switch number one, ground the bare wire to the box with a grounding clamp.
19. Attach the white wire to the single screw at one end of the switch after marking it with black tape.
20. Attach the red and black wires to the other two screws at the other end of the switch.
21. Attach the switch to the box with the screws provided.
22. Test the circuit.
23. Wire the circuit to the circuit breaker.
24. Install the switch covers.
25. Clean the area and put away the materials.

    *Note: When the switches have a green screw, you will need to bundle an 8" bare wire with a wire nut to ground both the switch and the box.

Instructor's Score or Approval _____

Job Sheet 32-7

Name _____ Date _____

Wiring Three-Way Switches with the Power Source at One Switch and the Lamp Holder Beyond the Switches

Objective

When you have completed this activity you will have demonstrated the ability to wire three-way switches with the lamp holder beyond the switches.

Tools and Materials

- Job sheet 32-7
- 2 three-way switches
- Lamp holder
- 2 wall boxes
- Ceiling box
- 14-3 with ground cable
- 14-2 with ground cable
- Grounding clamps
- Wire nuts
- Circuit tester
- Cross-point and regular screwdrivers

Procedure

1. Install the wall and ceiling boxes.
2. Install the 14-3 cable between the wall boxes.
3. Install the 14-2 cable between the entrance panel and the first wall box and between the second wall box and the ceiling box.
4. Strip 8″ of the cable ends.
5. Strip the ends of the wire.
6. At the first wall box install the three-way switch as in Job Sheet 32-5.
7. At the second wall box, cut an 8″ piece of bare wire and bundle it with the other two bare wires with a wire nut.
8. Connect the loose bare wire to the box with a grounding clamp.
9. Connect the two white wires with a wire nut.
10. Connect the red wire to one of the two screws at one end of the switch, and the black wire to the other screw.
11. Connect the black wire from the lamp holder to the single screw at the other end of the switch.
12. Install the switch with the screws provided.
13. At the ceiling box, ground the bare wire to the box with the grounding clip.
14. Attach the black wire to the brass screw and the white wire to the silver screw.
15. Install the lamp holder.
16. Correct the circuit at the circuit breaker.
17. Test the circuit.
18. Install the switch covers.
19. Clean the area and put away the materials.

Instructor's Score or Approval _____

Job Sheet 32-8

Name _____ Date _____

Wiring a Circuit with One or More Four-Way Switches

Objective

When you have completed this activity you will have demonstrated the ability to wire a circuit with one or more four-way switches.

Tools and Materials

Job sheet 32-8	Needle nose pliers
2 three-way switches	Cross-point and regular screwdrivers
Four-way switch	3 switch covers
Lamp holder	3 wall boxes
14-2 with ground cable	Ceiling box
14-3 with ground cable	Grounding clamps
Diagonal cutting pliers	Wire nuts

Procedure

1. Install the boxes.
2. Install the 14-2 cable from the entrance panel to the first wall box and from the third wall box to the ceiling box.
3. Install the 14-3 cable between the first and second, the second and third boxes.
4. Strip 8" of the cables.
5. Strip 5/8" of each wire.
6. At the first wall box, install a three-way switch as in Job Sheet 32-7.
7. At the second wall box, cut 8" of bare wire and clamp the three wires with a wire nut.
8. Ground the loose bare wire to the box.
9. Connect the two white wires with a wire nut.
10. Connect the red and black wires from the first switch to the two terminals on one end of the four-way switch.
11. Connect the red and black wires leading to the third box to the other two terminals on the other end of the switch.
12. Install the four-way switch with the screws provided.
13. Wire the three-way switch at the third box as in Job Sheet 32-7.
14. Wire the lamp holder as in Job Sheet 32-7.
15. Test the circuit.
16. Install the switch covers.
17. Connect the circuit at the circuit breaker.
18. Clean the area and put away the materials.

Instructor's Score or Approval _____

Unit 33 Electronics in Agriculture

More and more of our equipment items have one or more functions controlled electronically. Some understanding of the uses and care of electronic equipment will be useful to you.

Class Activity 33-1

Understanding Electronics

Shop Activity 33-2

Maintaining Electronic Equipment

Job Sheet 33-1

Name _____ Date _____

Understanding Electronics

Objective

Upon completing this activity you will have demonstrated an understanding of electronics.

Tools and Materials

Job sheet 33-1

Pen or pencil

Procedure

Complete the following worksheet.

1. _____ is said to be the most pervasive technology of the twentieth century.

2. Name five of our current electronic aids.
   _____
   _____

3. What is the unit of measurement for resistance?
   _____

4. Define conductance.
   _____
   _____

5. What is the unit of measurement for conductance?
   _____

6. How are small resistance modules identified?
   _____

7. What are the purposes of electrical multimeters?
   _____
   _____

8. What is the difference between an analog and a digital meter?
   _____
   _____

9. How is a resistance measurement completed?

10. How are amperes in a circuit measured?

11. How is the voltage in a circuit measured?

12. Define inductance.

13. What is the unit of measurement for inductance?

14. Define capacitance.

15. Define direct current.

16. Define alternating current.

17. Define cycle.

18. What is an oscilloscope?

19. What are semiconductors?

20. What are transistors?

21. What are thyristors?

22. What are integrated circuits?

23. Describe an optoelectrical device.

24. How are optoelectrical devices used?

25. Describe power suppliers.

26. What are amplifiers?

27. What are oscillators?

28. A digital computer processes information on the basis of a _____ number system.

29. T F Computer chips are the means of storing, processing, or printing many bits of information.

30. Floppy disks and computer tapes are the storage _____ for large amounts of data.

31. With the development of _____ and computer _____, computers became smaller and more reliable.

32. Name four common applications of microcomputers.

33. The _____ provides the mechanism for many agricultural applications.

34. Name one application of lasers in agriculture.

35. What part do electronics play in the modern planter?

36. Name three other agricultural applications.

37. What is a Global positioning system?

Instructor's Score or Approval _____

Job Sheet 33-2

Name _____ Date _____

Maintaining Electronic Equipment

Objective

When you have completed this activity you will have demonstrated the ability to maintain electronic equipment.

Tools and Materials

Job sheet 33-2

Pen or pencil

Procedure

Complete the following worksheet.

___ 1.	Read and follow	a.	dust and foreign materials
___ 2.	When it malfunctions	b.	a clean dry cloth
___ 3	Maintain a cool and dry	c.	the owner's manual
___ 4.	Wipe routinely with	d.	call a qualified technician
___ 5.	Do lubrication	e.	environment for most equipment
___ 6.	Keep the equipment free of	f.	as recommended by the manufacturer

Instructor's Score or Approval _____

Unit 34 Electric Motors, Drives, and Controls

Because electric motors are compact, economical, efficient, safe, and easy to control they are widely used in agriculture, homes, and industry. The understanding of their installation and control is a usable skill in agricultural mechanics.

Class Activity 34-1

Understanding Electric Motors, Drives, and Controls

Class Activity 34-2

Electric Motor Selection

Class Activity 34-3

Identifying Electrical Control Switches

Shop Activity 34-4

Maintaining Motors, Drives, and Controls

Job Sheet 34-1

Name _____ Date _____

Understanding Electric Motors, Drives, and Controls

Objective

Upon completing this activity you will have demonstrated an understanding of electric motors, drives, and controls.

Tools and Materials

Job sheet 34-1

Pen or pencil

Procedure

Complete the following worksheet.

1. The text suggests that electric motors are the _____ of modern technology.

2. Electric motors are the most common power source because they are:

   _____
   _____
   _____
   _____
   _____
   _____

3. Name the factors to be considered in selecting an electric motor.

   _____
   _____
   _____

4. What information is found on the motor nameplate?

   _____
   _____
   _____

5. A direct drive motor is connected _____ to the load.

6. What are other common power delivery systems?

   _____
   _____
   _____

7. How is pulley size computed?

   _____
   _____
   _____

8. What is the major advantage of a chain and sprocket delivery system over a belt drive?

   _____

9. How are electric motors controlled?
   _____
   _____
   _____

10. Fuses and circuit breakers are devices to prevent circuit _____

11. Motors are also frequently protected by built-in _____ devices.

12. What are the common type of switches used to control electric motors?
    _____
    _____
    _____

13. List the sensors and other switch actuators described. Give a use for each.
    _____
    _____
    _____
    _____
    _____

14. Describe how electric motors fit into a system such as a furnace.
    _____
    _____
    _____
    _____
    _____
    _____
    _____

15. When electric motors are installed or moved, care must be exercised in pulley and belt _____

Instructor's Score or Approval _____

Job Sheet 34-2

Name _____ Date _____

Electric Motor Selection

Objective

When you have completed this activity you will have demonstrated the ability to select electric motors.

Tools and Materials

    Job sheet 34-2

    Pen or pencil

    Cooper Text

Procedure

Complete the following worksheet.

In column I, indicate whether the motor is suited for an easy starting or hard starting load; in column II, indicate starting torque compared to running torque.

	Column I	*Column II*
Shaded-pole induction	____ 1. Easy start	____ a. 1/2 to 1 time
Split phase	____ 2. Hard start	____ b. 1 to 1 1/2 times
Permanent-split, capacitor-induction	____	____ c. 2 to 3 times
Soft-start	____	____ d. 3 to 4 times
Capacitor-start, induction-run	____	____ e. 3 1/2 to 4 1/2 times
Repulsion-start, induction-run	____	____ f. 4 times
Capacitor-start, capacitor-run	____	____
Repulsion-start, capacitor-run	____	____
Three-phase general purpose	____	____

Instructor's Score or Approval _____

> **Job Sheet 34-3**

Name _____ Date _____

Identifying Electrical Control Switches

Objective

When you have completed this activity you will have demonstrated the ability to identify electrical switch controls.

Tools and Materials

    Job sheet 34-3

    Pen or pencil

    Textbook

Procedure

Complete the checklist by matching the items in column I with those in column II.

	Column I		*Column II*
___	1. Float switch	a.	activates a switch after a preset time
___	2. Pilot light	b.	responds to temperature changes
___	3. Timer	c.	glows when the circuit is energized
___	4. Heat sensor	d.	activated by a float
___	5. Flow switch	e.	activated by flow
___	6. Pressure switch	f.	determines content
___	7. Gas analyzer	g.	reacts to pressure

Instructor's Score or Approval _____

Job Sheet 34-4

Name _____ Date _____

Maintaining Motors, Drives, and Controls

Objective

When you have completed this activity you will have demonstrated the ability to maintain motors, drives, and controls.

Tools and Materials

    Job sheet 34-4

    Motors with drives and controls

    Owner's manuals for the motors

    Cleaning rags and solvents

    Lubricants needed

Procedure

1. Maintain the motor by following Job Sheet 33-2.
2. Grease shafts or lubricate according to the owner's manual.
3. Clean chains of dirt and grit.
4. Oil chains as recommended.
5. Change gearbox lubricants as recommended.
6. Adjust belt tensions to the recommended deflection.
7. Check and adjust belt and pulley alignment.
8. Clean the area and put away the materials.

Instructor's Score or Approval _____

SECTION 12 PLUMBING, HYDRAULIC, AND PNEUMATIC SYSTEMS

Unit 35 Plumbing

Water is only one of the materials moved through tubular products. Plumbing is the term applied to system pipes and fixtures for water or gas in a building. In agricultural mechanics the term is applied to all work with pipe and fixtures.

Class Activity 35-1

Understanding Plumbing

Shop Activity 35-2

Connecting Plastic Pipe

Shop Activity 35-3

Repairing Faucets

Shop Activity 35-4

Doing a Plumbing Exercise

Job Sheet 35-1

Name _____ Date _____

Understanding Plumbing

Objective

Upon completion of this activity you will have demonstrated an understanding of plumbing.

Tools and Materials

   Job sheet 35-1

   Pen or pencil

Procedure

Complete the following worksheet.

1. What does the term plumbing mean in agricultural mechanics?

   _____

   _____

2. Define pipe. _____

3. What is a pipe fitting?

   _____

   _____

4. List the tools commonly used in plumbing.

   _____

   _____

   _____

5. How is pipe size commonly determined?

   _____

6. Describe galvanized pipe. _____

7. What is the caution in working with galvanized pipe and heat?

   _____

   _____

   _____

8. Heavy or double extra heavy pipe is usually used for construction instead of _____.

9. What are the two types of copper pipe?

   _____

   _____

10. How are these two types joined?

    _____

    _____

11. What do the designations K, L, or M refer to?

    _____

    _____

12. Which type is required if the pipe is to be buried?
    _____
    _____

13. What are the common plastic pipe materials in use today?
    _____
    _____
    _____

14. Why are plastic pipes most popular today?
    _____
    _____

15. T F Dripping or hammering may be corrected by tightening or replacing the faucet seal or washer.

16. What are some common problems with float valves?
    _____
    _____

17. Why should the septic system be placed far away from the well?
    _____
    _____
    _____

18. Why should septic tanks be pumped out at intervals?
    _____
    _____
    _____

Instructor's Score or Approval _____

Job Sheet 35-2

Name _____ Date _____

Connecting Plastic Pipe

Objective

When you have completed this activity you will have demonstrated the ability to connect plastic pipe.

Tools and Materials

    Job sheet 35-2

    PVC pipe, 1/2"

    PVC cleaner

    PVC cement

    Copper male adapter, 1/2"

    Galvanized male adapter, 1/2"

    Miter box

    Miter box saw

    Knife

    Rags

Procedure

1. Cut the PVC pipe to 12" in length using the miter box.
2. Remove the rough edges with the knife.
3. Clean the outside of the pipe ends and the inside of the fittings with a cloth.
4. Clean the pipe and fittings with the PVC cleaner.
5. Swab the ends of the pipe and the insides of the fittings with the PVC cement.
6. Insert the pipe into the fitting and give a quarter turn.
7. Hold them for a few seconds while the glue sets.
8. Clean the area and put away the materials.

Instructor's Score or Approval _____

Job Sheet 35-3

Name _____ Date _____

Repairing Faucets

Objective

When you have completed this activity you will have demonstrated the ability to repair a faucet.

Tools and Materials

Job sheet 35-3

A faucet with a leaking packing nut or drip

Screwdrivers

Adjustable end wrench

O-ring or packing

Faucet washers

Procedure

1. Tighten the packing nut and see if that stops the leak.
2. If the leak does not stop, shut off the water to the faucet.
3. Remove the faucet handle.
4. Remove the packing nut.
5. Replace the O-ring or packing and replace the packing nut.
6. Tighten the packing nut just enough to stop the leak.
7. If the faucet were dripping, remove the faucet stem and replace the faucet washer.
8. Tighten the washer snugly without distortion.
9. Reassemble the faucet.
10. Test for leaks.
11. Clean the area and put away the materials.

Instructor's Score or Approval _____

Job Sheet 35-4

Name _____ Date _____

Doing a Plumbing Exercise

Objective

When you have completed this activity you will have demonstrated the ability to do a plumbing exercise using a number of different materials.

Tools and Materials

Job sheet 35-4

PVC pipe and fixtures from Job Sheet 35-2

Polyethylene pipe, 1/2"

1/2" hose adapter, female (galv.)

1/2" tee (galv.)

1/2" coupling (galv.)

1/2" elbow (galv.)

1/2" sillcock, female (galv.)

1/2" x 4" nipple (galv)

1/2" x 6" nipple (galv.)

Two 1/2" male adapter inserts

Two 1/2" stainless steel pipe clamps

1/2" adapting elbow galvanized to copper

1/2" copper tee

1/2" copper union

1/2" copper adapter, male

15" type L copper pipe

Flux, solvents, and tools necessary to complete the exercise

Procedure

1. Using the pipe compound, assemble the hose adapter to one end of the galvanized tee.
2. Assemble the 4" nipple to the other end of the tee.
3. Assemble one of the male inserts to the tee.
4. Attach the 1/2" coupling to the 4" nipple.
5. Assemble the 6" nipple to the coupling.
6. Sweat the male adapter to one end of the copper tee.
7. Cut 6" of copper tubing and sweat it to the other end of the tee.
8. Sweat the female adapter to the tee.
9. Sweat the male part of the union to the 6" pipe.
10. Sweat the male part of the union to a copper pipe cut to equal the measurement of the galvanized assembly.
11. Slip the compression ring on the pipe and sweat the adapting elbow on the copper pipe.

12. Assemble the other male insert to the adapting elbow.
13. Assemble the elbow and the polyethylene pipe with the stainless steel clamp.
14. Assemble the 12" polyethylene pipe to the galvanized tee adapter with the other stainless steel clamp.
15. Assemble the male adapter in the copper tee and the sillcock.
16. Assemble the PVC pipe to the galvanized tee adapter.
17. Assemble the copper tee to the unit from step 16.
18. Align and assemble the copper union.
19. Attach the hose and check for leaks.
20. Clean the area and put away the materials.

Instructor's Score or Approval _____

Unit 36 Irrigation Technology

Irrigation has been used by many civilizations over the centuries. This practice has enabled higher production and reclamation of otherwise less-productive land.

Class Activity 36-1

Identifying Irrigation Systems

Class Activity 36-2

Understanding Irrigation

Job Sheet 36-1

Name _____ Date _____

Identifying Irrigation Systems

Objective

When you have completed this activity you will have demonstrated the ability to identify irrigation systems.

Tools and Materials

Job sheet 36-1

Pen or pencil

Procedure

Complete the following worksheet.

___ 1. Surface   a. simulated rain

___ 2. Subirrigation   b. by gravity

___ 3. Trickle   c. by using subsoil

___ 4. Sprinkler   d. by using small emitters

Instructor's Score or Approval _____

Job Sheet 36-2

Name _____ Date _____

Understanding Irrigation

Objective

Upon completion of this activity you will have demonstrated an understanding of irrigation.

Tools and Materials

Job sheet 36-2

Pen or pencil

Procedure

Complete the following worksheet.

1. Define irrigation.

   _____
   _____

2. What are the benefits of irrigation?

   _____
   _____
   _____
   _____
   _____

3. List the 12 questions to consider in making an irrigation decision.

   _____
   _____
   _____
   _____
   _____
   _____

4. An inch of water on an acre of soil requires _____ gallons.

5. Define seasonal demand.

   _____
   _____
   _____

6. How do daily water requirements for alfalfa grown in Virginia compare to southern Texas?

   _____
   _____

7. Explain the term peak-use.

   _____
   _____
   _____

8. T F Before starting an irrigation system, one should check local water-use laws.
9. T F Irrigation wells and pumps must be designed to fit a given installation.
10. T F Planning irrigation systems requires no knowledge of plant growth.
11. What are moisture sensors?

_____
_____
_____

Instructor's Score or Approval _____

Unit 37 Hydraulic, Pneumatic, and Robotic Power

Fluid power is widely used in many industries as well as agriculture. Liquids and air, or gases, are both considered fluids. The difference is the compressibility of the gases.

Class Activity 37-1

Understanding Pneumatic, Hydraulic, and Robotic Power

Class Activity 37-2

Identifying Fluid Additives

Class Activity 37-3

Identifying Hydraulic System Maintenance Items

Class Activity 37-4

Identifying Robotic Work Areas

Job Sheet 37-1

Name _____ Date _____

Understanding Pneumatic, Hydraulic, and Robotic Power

Objective
Upon completing this activity you will have demonstrated an understanding of pneumatics, hydraulics, and robotic power.

Tools and Materials
Job sheet 37-1   Pen or pencil

Procedure
Complete the following worksheet.

1. Define hydraulics.
   _____
   _____

2. Define pneumatics.
   _____
   _____

3. These two power sources are known as _____ power.

4. T F A log splitter would be a use of pneumatic power.

5. T F The log splitter would be an example of fluid power.

6. Define force.
   _____
   _____

7. Define pressure.
   _____
   _____

8. T F Pressure in liquid containers is exerted equally on all surfaces.

9. T F When equal pressure is applied to cylinder A, the resulting force will be greater in the larger of cylinders B, or C.

10. T F Boyle's Law describes why basketballs bounce.

11. T F As the diameter of a pipe decreases, the pressure increases.

12. What is the meaning of the SAE rating of 10W40?
    _____
    _____
    _____

13. What is the meaning of an API rating?
    _____
    _____
    _____

14. List the components of a hydraulic system.
    _____
    _____
    _____

15. Name the three types of hydraulic pumps.
    _____
    _____
    _____

16. Describe a fluid coupling.
    _____
    _____
    _____

17. List the desirable characteristics of hydraulic hoses.
    _____
    _____
    _____

18. Describe the function of a pressure regulator.
    _____
    _____
    _____

19. Describe a check or direction-control valve.
    _____
    _____
    _____

20. Describe an equipment control valve.
    _____
    _____
    _____

21. Why are filtering systems needed in hydraulic systems?
    _____
    _____
    _____

22. What is a full-flow filtering system?
    _____
    _____

23. What is a bypass filtering system?
    _____
    _____

24. Describe a hydraulic cylinder.
   _____
   _____
   _____

25. Describe a hydraulic motor.
   _____
   _____
   _____

26. Describe the maintenance needed by the following components.
   Fluid level _____
   Fluid filters _____
   Quick couplings _____
   _____

27. What are the signals of a needed hydraulic fluid change?
   _____
   _____
   _____

28. T F Pneumatic system components are often similar to hydraulic.

29. The greatest difference is their
   _____

30. Define robotics.
   _____
   _____

31. Robots are distinguished from other machines because they
   _____
   _____
   _____
   _____

32. Robots are well suited to tasks that involve
   _____
   _____
   _____

33. A robot's motions are governed by _____ and _____ actions called _____.

34. What is the biggest advantage of electricity for powering robots?
   _____
   _____

Instructor's Score or Approval _____

Job Sheet 37-2

Name _____ Date _____

Identifying Fluid Additives

Objective

When you have completed this activity you will have demonstrated the ability to identify fluid additives.

Tools and Materials

Job sheet 37-2

Pen or pencil

Procedure

Complete the checklist and review it with your instructor.

___ 1. Viscosity improver   a. to prevent oil breakdown

___ 2. Antifoam   b. helps polish moving parts

___ 3. Corrosion inhibitor   c. reduces or prevents rust

___ 4. Rust inhibitor   d. provides stability with temperature change

___ 5. Antiscuff   e. reduces foaming

___ 6. Extreme-pressure resistor   f. reduces or prevents corrosion

Instructor's Score or Approval _____

Job Sheet 37-3

Name _____ Date _____

Identifying Hydraulic System Maintenance Items

Objective

When you have completed this activity you will have demonstrated the ability to identify hydraulic system maintenance items.

Tools and Materials

Job sheet 37-3

Pen or pencil

Procedure

Complete the checklist and review it with your instructor.

___ 1. Fluid level
___ 2. Fluid filters
___ 3. Quick-couplings
___ 4. Fluid changes

a. drain and refill at the manufacturer's time interval
b. keep clean by capping when not in use
c. checked frequently and maintained
d. unscrew and replace at specified intervals

Instructor's Score or Approval _____

Job Sheet 37-4

Name _____ Date _____

Identifying Robotic Work Areas

Objective

When you have completed this activity you will have demonstrated the ability to identify the robotic work areas.

Tools and Materials

Job sheet 37-4

Pen or pencil

Procedure

Complete the following worksheet.

___ 1. Cartesian   a. three rotational axes

___ 2. Cylindrical   b. two rotational and one translational axes

___ 3. Hollow sphere   c. box-like work area

___ 4. Solid sphere   d. one rotational and two translational axes

Instructor's Score or Approval _____

SECTION 13 CONCRETE AND MASONRY

Unit 38 Concrete and Masonry

Concrete and masonry materials are very widely used. A knowledge of concrete construction is a basic skill in agricultural mechanics because of its many uses and the flexibility of its forming and placement.

Class Activity 38-1

Understanding Concrete and Masonry

Shop Activity 38-2

Testing Sand for Silt or Clay

Class Activity 38-3

Estimating Materials

Shop Activity 38-4

Preparing Forms

Shop Activity 38-5

Laying a Concrete Block Corner

Job Sheet 38-1

Name _____ Date _____

Understanding Concrete and Masonry

Objective

Upon completing this activity you will have demonstrated an understanding of concrete and masonry.

Tools and Materials

Job sheet 38-1

Pen or pencil

Procedure

Complete the following worksheet.

1. Define masonry.

   _____
   _____

2. Define cement.

   _____
   _____

3. Define concrete.

   _____
   _____

4. What is mortar?

   _____
   _____

5. What are the advantages of concrete as a building material?

   _____
   _____
   _____
   _____
   _____
   _____

6. Why does quality concrete require the use of graded aggregates?

   _____
   _____

7. Why should sand be tested for silt or clay?

   _____
   _____
   _____

8. What are the characteristics of a workable mix?

   _____
   _____
   _____
   _____

9. What are concrete forms?

   _____
   _____

10. What materials are usually used in form construction?

    _____
    _____

11. What is a construction joint? A control joint?

    _____
    _____
    _____

12. Why and how is concrete reinforced?

    _____
    _____
    _____

13. Identify the concrete finishing tools pictured below.

Figure 38-1. Concrete Tools

14. How soon does concrete start to harden after mixing is stopped?

15. What is screeding? Why is it needed?

16. What is floating? Why is it done?

17. What is air-entrained concrete? Why is it used?

18. What is a broom finish? Why is it used?

19. What is curing? Why is moisture important?

20. Identify the common types of blocks shown below.

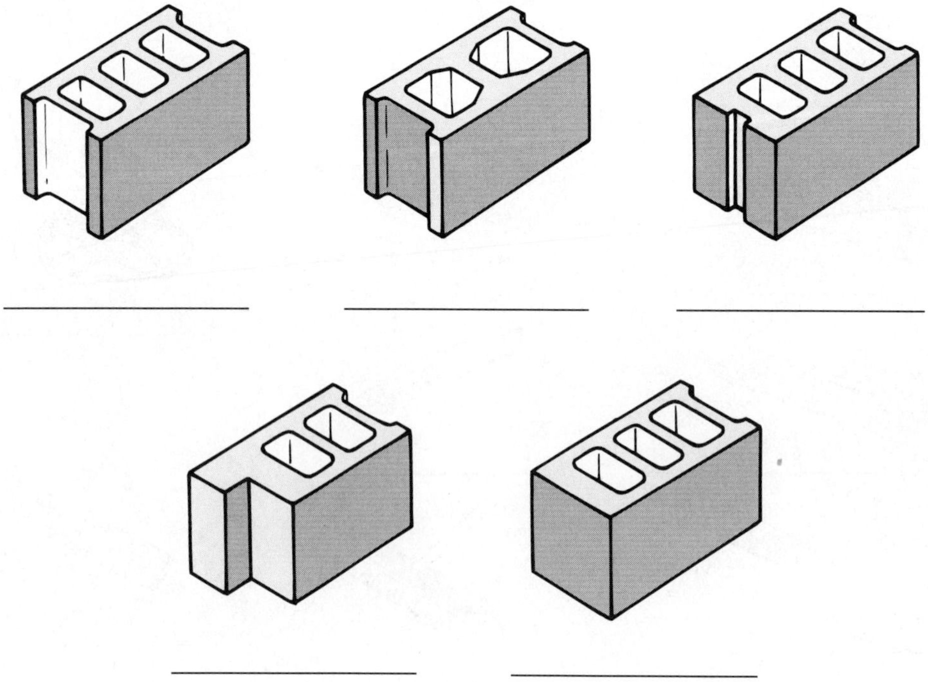

Figure 38-2. Types of Concrete Blocks

21. What is the nominal size of a concrete block?

22. How many blocks will be needed for a wall 36′ long and 8′ high with a 2′ 8″ x 6′ 8″ door and a 32″ x 48″ window?

23. What is a footer? How deep should it be placed?

24. Why must the wetness of the mortar be carefully controlled?

25. How are blocks installed straight, level, and plumb?

Instructor's Score or Approval _____

Job Sheet 38-2

Name _____ Date _____

Testing Sand for Silt or Clay

Objective

When you have completed this activity you will have demonstrated the ability to test sand for silt or clay.

Tools and Materials

    Job sheet 38-2

    Quart fruit jar

    Sample of sand to be tested

    Water

    Ruler

Procedure

1. Place 2" of sand in the quart jar.
2. Add water to 3/4 full.
3. Cap the jar and shake vigorously for one minute.
4. Shake sideways to level the sand.
5. Set the jar aside for one hour.
6. Measure the silt at the top of the sand.
7. If the layer of silt is over 1/8", reject the sand or wash it.
8. If the layer is under 1/8", let it set overnight to check the clay content.
9. If the layer is less than 1/8", it may be used, and if it is over it is rejected or washed.
10. Clean the area and put away the materials.

Instructor's Score or Approval _____

Job Sheet 38-3

Name _____ Date _____

Estimating Materials

Objective

When you have completed this activity you will have demonstrated the ability to estimate materials needed for a concrete job.

Tools and Materials

    Job sheet 38-3

    Pen or pencil

Procedure

Determine the materials for a sidewalk 4' wide x 40' long x 4" thick. Use a 1, 2 1/2, 3 1/2 mix (see text Figure 38-5).

1. Compute the cubic feet in the sidewalk   _____ cubic feet
2. Compute the cubic yards in the sidewalk   _____ cubic yards
3. Compute the amount of cement needed   _____ pounds or cubic feet
4. Compute the amount of fine aggregate needed   _____ pounds or cubic feet
5. Compute the amount of coarse aggregate needed   _____ pounds or cubic feet

    Optional; Compute the cost of the materials.

Instructor's Score or Approval _____

Job Sheet 38-4

Name _____ Date _____

Preparing Forms

Objective

When you have completed this activity you will have demonstrated the ability to construct concrete forms.

Tools and Materials

- Job sheet 38-4
- 2" x 4" material for forms and stakes
- Sledge hammer
- Claw hammer
- 8d nails
- Level
- Table saw
- Carpenter's square
- Shovel
- Handsaw
- Reasonably level area for construction

Procedure

1. Locate one corner for the size form your instructor has chosen.
2. Cut the 2" x 4" ends and sides needed.
3. Cut and sharpen a stake for each corner plus one for every 3 feet of each side and end.
4. Set one side of the form and stake it every 3 feet.
5. Level the side and nail through the form into the stakes.
6. Establish a square corner with the carpenter's square and stake the end.
7. Level the end to the side and nail into the stakes.
8. Do the other side and end in the same manner.
9. Saw any stakes above the form level with the forms.
10. Clean the area and put away the materials.

    Note: If concrete were to be poured in these forms, the additional steps of leveling the surface within the forms to the depth of the forms would be accomplished.

Instructor's Score or Approval _____

Job Sheet 38-5

Name _____ Date _____

Laying a Concrete Block Corner

Objective

When you have completed this activity you will have demonstrated the ability to lay a concrete block corner.

Tools and Materials

Job sheet 38-5	Mixing hoe
Mortar box	Chalk line
Mortar mix	Level with a plumb
Water	Water
21 stretcher blocks	Carpenter's square
7 corner blocks	Story pole or stakes
Mason's trowel	

Procedure

1. Establish the corner with the chalk line and square.
2. Mix the mortar to a stiff consistency.
3. Spread a layer of mortar on the footer.
4. Set a corner block precisely on the corner and level it to the chalk line.
5. Set 3 stretcher blocks on end and apply mortar to the ears with a wiping motion.
6. Set the mortared ends against the first block and use the handle of the trowel to level and plumb them to the left of the first block.
7. Spread a mortar bed to the right of the corner block and repeat step 6.
8. Move the chalklines up 8 inches.
9. Place a mortar bed across the first corner block and the stretcher block beside it.
10. Set the next corner block across these two blocks and level and plumb it to the chalk line.
11. Complete the mortar bed and set stretcher blocks as in steps 5 and 6.
12. Repeat on the other side of the corner.
13. Repeat steps 8 to 12 until the corner is complete, remembering that each course uses one less block.
14. Cut the excess mortar off with the trowel.
15. After the mortar dries a bit, finish the joint with a block or jointer.
16. Remove and clean the blocks before the mortar hardens completely.
17. Clean the area and put away the materials.

Instructor's Score or Approval _____

SECTION 14 AGRICULTURAL STRUCTURES

Unit 39 Planning and Constructing Agricultural Structures

Many skills are needed in constructing and maintaining agricultural buildings or structures. These same skills are widely used in other industries.

Class Activity 39-1

Understanding Agricultural Structures

Class Activity 39-2

Identifying Veneer Grades

Shop Activity 39-3

Laying Out a Building

Shop Activity 39-4

Laying Out a Stairs

Shop Activity 39-5

Laying Out a Rafter

Shop Activity 39-6

Tying a Square Knot

Shop Activity 39-7

Tying a Bowline Knot

Shop Activity 39-8

Tying a Clove Hitch

Shop Activity 39-9

Tying Two Half Hitches

Shop Activity 39-10

Tying a Timber Hitch

Shop Activity 39-11

Tying a Sheet Bend

Job Sheet 39-1

Name _____ Date _____

Understanding Agricultural Structures

Objective

Upon completing this activity you will have demonstrated an understanding of agricultural structures.

Tools and Materials

Job sheet 39-1   Pen or pencil

Procedure

Complete the following worksheet.

1. Planning is important because it saves both _____
   and _____.

2. List the principles of farmstead layout.
   _____
   _____

3. Name a farm building for which professional help in design is recommended.
   _____

4. A farm family might plan a _____
   _____

5. Define rafter. _____
   _____

6. Define girder. _____
   _____

7. Define a truss. _____
   _____

8. Buildings are classified by the _____ or the type of
   _____

9. A pole building is one supported by _____.

10. The most common materials used for agricultural buildings are _____
    _____

11. Roofing and siding of agricultural buildings may be of _____
    or _____.

12. What is OSB?

13. Why is OSB important? _____

14. Insulation materials commonly used include _____

15. When livestock buildings are insulated _____
    _____ is very important.

Instructor's Score or Approval _____

Job Sheet 39-2

Name _____ Date _____

Identifying Veneer Grades

Objective

When you have completed this activity you will have demonstrated the ability to identify veneer lumber grades.

Tools and Materials

Job sheet 39-2

Pen

Procedure

Match the items in column I with those in column II.

Column I

___ 1. A
___ 2. B
___ 3. C
___ 4. C plugged
___ 5. D

Column II

a. Improved C veneer; splits limited to 1/8" and knots and borer holes limited to 1/4" x 1/2"

b. Tight knots to 1 1/2" or knotholes to 1" across grain

c. Knots and knotholes to 2 1/2" across grain

d. Smooth and paintable with limited repairs permitted with the grain

e. Solid surface with shims and circular repair plugs and tight knots to 1"

Instructor's Score or Approval _____

Job Sheet 39-3

Name _____ Date _____

Laying Out a Building

Objective

When you have completed this activity you will have demonstrated the ability to lay out a building.

(**Note:** This is similar to, but more advanced than, the procedure in shop activity 8-2.)

Tools and Materials

Job sheet 39-3

Line level or transit

16 2" x 2" x 16" stakes

8 1" x 4" x 3' boards

Hammer

Nails

Chalk lines

Sledge hammer

Procedure

1. Locate one corner of the proposed building and set one stake, driving a nail into the top of the stake as a reference point.
2. Measure the length of the building in the desired direction from this stake and set the second stake level with the first, using the line level or transit.
3. Drive a nail in the top of the stake exactly at the length of the building from the nail in the first stake.
4. Measure the width of the building perpendicular to the length of the building established in steps 2 and 3 (use the 3-4-5 right triangle method).
5. Set the third temporary stake and level it with the first stake.
6. Drive a nail in the stake at the exact length of the end of the building.
7. Measure the length of the building parallel to the first side laid out.
8. Measure the second end of the building from the second stake and set a stake where the line crosses the line in step 7, leveling the stake with the other three.
9. Drive a nail at the exact distance from stakes 2 and 3.
10. Measure the distance from this nail to the first, and from the second to the third.
11. If the distances are unequal, shift the third and fourth stakes and nails 3 or 4 and 2 as needed.
12. With the stakes leveled and the nails at the proper distances, install batter boards a foot beyond the stakes.
13. Level the batter boards with the stakes.
14. Using saw kerfs or nails, establish chalk lines that parallel the nails on the stakes.
15. Remove the stakes and use the chalk lines to reference the corners.
16. Clean the area and put away the materials.

Instructor's Score or Approval _____

Job Sheet 39-4

Name _____ Date _____

Laying Out a Stairs

Objective

When you have completed this activity you will have demonstrated the ability to lay out a stairs.

Tools and Materials

Job sheet 39-4

2" x 10" x 10'

Carpenter's square

Pencil

2 sawhorses

2' bench rule

2 small C-clamps

Procedure

1. Lay the 2" x 10" on the sawhorses.
2. Compute the risers and treads assuming a 60" rise.
3. Clamp the bench rule on the carpenter's square with the tread measure on the blade and the riser measure on the tongue.
4. Place the carpenter's square on the 2" x 10" with the tongue at the left end.
5. Adjust the square along the edge until the full riser minus the thickness of the tread can be marked at the end of the board.
6. Reverse the square and place the blade at this mark, then mark the tread and riser cuts.
7. Slide the square to the riser mark and mark the next cuts.
8. Repeat step 7 until all cuts are marked.
9. Clean the area and put away the materials.

Instructor's Score or Approval _____

Job Sheet 39-5

Name _____ Date _____

Laying Out a Rafter

Objective

When you have completed this activity you will have demonstrated the ability to lay out a rafter.

Tools and Materials

Job sheet 39-5

Carpenter's square

2" x 4" x 10'

Pencil

2 saw horses

2' bench rule

Procedure

Assume a run of 6' and a rise of 2' with an overhang of 6".

1. Lay the 2" x 4" on the sawhorses.
2. Lay out a marking line 2" from the bottom edge of the raftor the length of the rafter.
3. Lay the square on the rafter with the tongue nearest the end and put the 4" on the marking line and 12" on the blade on the marking line.
4. Adjust the square until it is within 1/4" of the end and mark the plumb cut.
5. With the square on the plumb line, adjust the square until 6" on the blade intersects the marking line, and pencil a mark for the point of the bird's mouth.
6. Mark a plumb line at this point to the bottom of the rafter for the bird's mouth.
7. Reverse the square and square the bird's mouth by putting the 4" on the tongue and the 12" on the blade from the pencil mark.
8. In the text see Figure 39-28. Follow the procedure.
9. Calculate the length of the rafter and measure the distance along the marking line from the point of the bird's mouth.
10. Mark a plumb cut parallel to the first or overhang cut mark.
11. Clean the area and put away the materials.

Note: You could step off the rafter by using the technique for laying out a stairs.

Instructor's Score or Approval _____

Job Sheet 39-6

Name _____ Date _____

Tying a Square Knot

Objective

When you have completed this activity you will have demonstrated the ability to tie a square knot.

Tools and Materials

Job sheet 39-6

2′ of rope

Procedure

1. Hold the ends of the rope in your left hand with the ends toward the outside.
2. Bring the right end under and the left end under and loop the ends back to the center.
3. With the end now on the left in front, wrap the rope around and pull it through the loop formed.
4. With the other end, bring the rope over the first and pull it through the loop formed.
5. Check that the result looks like text Figure 39-31.
6. Practice the knot until it is easy to complete.
7. Move on to the next exercise.

Instructor's Score or Approval _____

Job Sheet 39-7

Name _____ Date _____

Tying a Bowline Knot

Objective

When you have completed this activity you will have demonstrated the ability to tie a bowline knot.

Tools and Materials

Job sheet 39-7

Rope used in Job sheet 39-6

Procedure

1. Hold the rope in the middle with the left hand.
2. Form a 2" to 3" circle with the right hand and bring the rope back over the top of the end in the left hand.
3. Form a larger circle and bring the right end up through the first circle.
4. Take this end under the rope in the left hand and around the rope.
5. Bring the end down through the first circle and tighten.
6. Check to see that it looks like text Figure 39-31.
7. Practice the knot several times.
8. Continue to the next exercise.

Instructor's Score or Approval _____

Job Sheet 39-8

Name _____ Date _____

Tying a Clove Hitch

Objective

When you have completed this activity you will have demonstrated the ability to tie a clove hitch.

Tools and Materials

    Job sheet 39-8

    Rope used in the previous activity

    Vise

    Short handle

Procedure

1. Place the short handle in the vise so it extends a foot or more vertically.
2. Form a 3" circle with the left end on top.
3. Place this circle over the short handle.
4. Form another circle with the right end with the right end underneath.
5. Drop the circle over the handle and tighten (see text Figure 39-31).
6. Practice the knot several times.
7. Proceed to the next exercise.

Instructor's Score or Approval _____

Job Sheet 39-9

Name _____ Date _____

Tying Two Half Hitches

Objective

When you have completed this activity you will have demonstrated the ability to tie two half hitches.

Tools and Materials

Job sheet 39-9

The rope used for the previous exercises

Vise

Short handle used for the previous exercise

Procedure

1. Set the short handle in the vise.
2. With one end of the rope in the left hand, pass the right end around the handle from the back.
3. Pass the rope over the end in the left hand and bring the end around and up through the loop around the handle.
4. Loop the rope around the part in the left hand again and pull it up through the resulting loop.
5. Tighten the ends.
6. Compare it to text Figure 39-31.
7. Practice the knot several times.
8. Continue to the next exercise.

Instructor's Score or Approval _____

Job Sheet 39-10

Name _____ Date _____

Tying a Timber Hitch

Objective

When you have completed this activity you will have demonstrated the ability to tie a timber hitch.

Tools and Materials

Job sheet 39-10

The rope used in the previous exercise

Vise

Short handle

Procedure

1. Set the handle in the vise.
2. Wrap the rope around the handle loosely in a clockwise direction below the other end.
3. Hold this rope up along the handle and bring the other end around and under itself twice and tighten.
4. With the upper end, form a second loop around the handle in a counterclockwise direction and pull it underneath the start of the loop.
5. Compare the result to text Figure 39-31.
6. Practice the knot several times.
7. Proceed to the next exercise.

Instructor's Score or Approval _____

Job Sheet 39-11

Name _____ Date _____

Tying a Sheet Bend

Objective

When you have completed this activity you will have demonstrated the ability to tie two different sized ropes together.

Tools and Materials

  Job sheet 39-11

  2′ of a smaller sized rope

  Rope used in the previous exercises

Procedure

1. Form a loop in the larger rope and hold the two ends in your left hand.
2. Pass one end of the smaller rope through the loop and around both sides of the loop.
3. Pass the end of the smaller rope under itself as it comes through the loop.
4. Compare this knot to text Figure 39-31.
5. Practice the knot several times.
6. Clean the area and put away the materials.

Instructor's Score or Approval _____

Unit 40 Aquaculture, Greenhouse, and Hydroponics Structures

Aquaculture and hydroponics production have joined greenhouse production as growing areas of agricultural enterprises. These activities take specialized structures.

Class Activity 40-1

Understanding Aquaculture, Greenhouse, and Hydroponics Structures

Class Activity 40-2

Identifying Aquaculture Systems

Class Activity 40-3

Identifying Greenhouse Types

Class Activity 40-4

Identifying Hydroponics Systems

Job Sheet 40-1

Name _____ Date _____

Understanding Aquaculture, Greenhouse, and Hydroponics Structures

Objective

Upon completing this activity you will have demonstrated an understanding of aquaculture, greenhouse, and hydroponics structures.

Tools and Materials

Job sheet 40-1

Pen or pencil

Procedure

Complete the following worksheet.

1. Describe an open aquaculture system.

   _____
   _____
   _____

2. Describe a semi-open system.

   _____
   _____
   _____

3. What additional actions must the grower take in using a semi-open system?

   _____
   _____
   _____
   _____
   _____

4. What are the two ways oxygen is added to the system?

   _____
   _____

5. Where are closed systems used?

   _____
   _____

6. What are the major components of a closed system?

   _____
   _____
   _____

7. What is the advantage of a round tank?

   _____
   _____

8. What is the advantage of a rectangular tank?

9. What is the advantage of a plastic liner?

10. What is the disadvantage of a plastic liner?

11. Auxiliary equipment capacity is based on fish _____ in the system.

12. Describe a greenhouse.

13. What are commonly used greenhouse covering materials?

14. In summer, the greenhouse must be _____ or _____ to control the temperature.

15. Greenhouse location should provide full _____ to the sun.

16. Wintertime heating requirements are determined by

17. Define hydroponics.

18. What are the components of a hydroponics growing system?

19. T F The hydroponics system produces the highest production of any plant culture system.
20. T F In the hydroponics system, very little supervision is required.

Instructor's Score or Approval _____

Job Sheet 40-2

Name _____ Date _____

Identifying Aquaculture Systems

Objective

When you have completed this activity you will have demonstrated the ability to identify aquaculture systems.

Tools and Materials

Job sheet 40-2

Pen or pencil

Procedure

Match the items in column I with those in column II.

	Column I		Column II
___ 1.	Open	a.	recirculating system
___ 2.	Semi-open system	b.	small density
___ 3.	Closed	c.	pumps and flushing

Instructor's Score or Approval _____

Lab Manual

Job Sheet 40-3

Name _____ Date _____

Identifying Greenhouse Types

Objective

When you have completed this activity you will have demonstrated the ability to identify greenhouse types.

Tools and Materials

Job sheet 40-3

Pen or pencil

Procedure

___ 1. Gable   a. rounded top

___ 2. Sawtooth   b. single-peaked

___ 3. Quonset   c. irregular top line

___ 4. Lean-to   d. added on to another building

Instructor's Score or Approval _____

Job Sheet 40-4

Name _____ Date _____

Identifying Hydroponics Systems

Objective

When you have completed this activity you will have demonstrated the ability to identify hydroponics systems.

Tools and Materials

Job sheet 40-4

Pen or pencil

Procedure

Match the items in column I with those in column II.

Column I *Column II*

___ 1. Bag culture   a. in continuously circulating nutrient solutions

___ 2. Trough culture   b. in bags with emitters

___ 3. Tube culture   c. in intermittent nutrient circulation

Instructor's Score or Approval _____

NOTES

NOTES

NOTES